Herefordshire Bricks & Brickmakers

Herefordshire Bricks & Brickmakers

by

Edwin Davey & Rebecca Roseff

Logaston Press

LOGASTON PRESS
Little Logaston Woonton Almeley
Herefordshire HR3 6QH
logastonpress.co.uk

First published by Logaston Press 2007

ISBN 978 1 904396 70 3

Set in Times New Roman by Logaston Press
and printed in Great Britain by
Bell & Bain Ltd., Glasgow

Contents

Abbreviations and Measurements used

HSM	Herefordshire Sites and Monuments Record (www.smr.herefordshire.gov.uk)
HRO	Hereford Record Office
PRO	Public Record Office
£1	240d, 12d to 1 shilling, 20s to £1
1 inch	25.4 mm, 12 inches to 1 foot = 30.48 cm
1 metre	3 feet 3.4 inches, or 3.281 feet
1 lb	0.454 kg, 1 kg = 2 lbs 3.2 oz or 2.2 lbs

Introduction and Acknowledgments

Bricks were the most common building material of 19th- and 20th-century Herefordshire. Up until the 1930s most were made locally either in large brickworks based in Hereford, Ledbury and Bromyard or from smaller kilns working near to the site. This local economy changed dramatically just before the Second World War as brick became one of the first major industries to succumb to economy of scale. Before 1937 nearly all houses in Herefordshire were made of local brick, after 1937 nearly all houses were made of imported brick. Most of these imported bricks were 'London' bricks, a much lighter, yellower brick, quite different to the heavy, hard, red home-made product. Their use and the different brick laying techniques that began about the same time has influenced the whole character of Hereford.

About 350 brickmaking sites are listed in this book. These are the main ones operating in the 19th and 20th centuries, but there would have been many more than this, for kilns were often short lived affairs. Little is known about the brickworks, and this is true even for the large factories. Few photographs exist, hardly any structures and we have only been able to find one person who remembers working in a brickworks in Herefordshire. There is much waiting to be discovered and to some extent this book is about prompting memories and inspiring people to come forward with their knowledge. If you have information about brickworks in Herefordshire please contact the authors, Rebeccaroseff@aol.com or Edwin Davey at Tel: 01432 880312.

This book gives an outline history of brickmaking in Herefordshire, together with a gazetteer of Herefordshire brickmaking sites. Bricks are heavy and the process of making them uses bulky raw materials. Comparatively large brickworks existed before canals or railways and it is intriguing to think how people managed so well using horse and people power. In most cases brickworks were located directly on the 'clay' source they used, but there must have been a network of tracks to bring in fuel and to move the finished product. Brickmaking was directly connected with wood coppicing and/or trade with coal producing regions in the West Midlands, South Wales and the Forest of Dean. It relied on people willing to work hard for comparatively low wages in a seasonal fashion.

Brickmaking was a craft learnt through practice and from other people, and we have almost lost the knowledge of how it was done. Robert Simpson and Alisdair Compton have researched brickmaking in the Welsh borders for many years and are re-learning the skills. Their work can be seen at Acton Scott Working Farm Museum, and much of the information given below has derived from talking with them and we are very grateful for their help and inspiration. Our editor Andy Johnson re-arranged the book and guided us through to publication. David Whitehead gave us access to his card index of brickmakers of the 16th to 18th centuries, the result of years of research. We have not done his work justice here. Alan Blake, builder, and Wilf Slade, bricklayer, of Hereford gave us many insights into the working and different qualities of Herefordshire brick. Terry Smith, of the Museum of London, told us about London bricks. Eric Conway of Baggeridge Bricks showed us around the Waresley Brickworks in Hartlebury and we learnt an enormous amount from his great knowledge of the trade. Judy Stevenson, Andy Boucher, Kath Crookes and Alan Vince told us about medieval brick, Duncan James about timber and housing in Herefordshire, David Smith about gypsies and travelling in Herefordshire. Marge Feryok, Martin Hudson and Steve Rexworthy brought interesting bricks to our attention. Finally Gerald Dawe and Joy Davey carried, photographed, became interested in and generally put up with bricks for several years and we are especially grateful to them.

1 Traditional Methods of Making Brick

Bricks are commonly said to be made of clay—although this is not strictly true, they are made of clay, silt, sand and stones in differing proportions—and usually fired, but not always so. Brick is an excellent building material, being a convenient size to hold, retaining but resistant to heat, keeping out noise, being strong and waterproof and having a pleasing, earthy colour. In addition, brick can last for literally thousands of years.

Clays are minute (<0.002 mm), platy shaped alumino-silicate minerals that are ionically active which means they take up water (and other molecules) chemically and thereby become plastic. When fired above 500°C they change their structure, losing their ionic attraction and becoming inert, meaning they cannot be changed by chemical means under normal conditions. In contrast silts and sands are inert particles of broken down rock differentiated by size (silt is 0.002 mm to 0.06 mm; sand 0.06 mm to 2 mm). Too much clay in a brick means it will be hard to work and expensive to fire as it holds too much water. Too little clay, means it will be less solid and firm. The ideal 'brickearth' for hand made bricks will contain about 25% clay with the remaining portion being mostly silt or fine sand. This makes a solid, firm block, not too hard to work. With modern machinery bricks can be made with higher clay content which make them harder, heavier and finer textured. This high clay content would have been difficult to work by hand in the past.

Builders classify bricks according to their load bearing qualities and appearance. A well fired brick with comparatively high clay content will take a load of 20 tonnes/0.1 sq. m (20 tons to the sq. ft), a partially fired brick, or those with low clay content have a lower load bearing rating and can only be used to make free standing structures such as garden walls.

Brick courses are fixed with mortar. A good mortar should be similar in strength and porosity to the brick it binds and be easy to work. It should never be harder than the brick as it will wear at a different rate and may lead to stresses and strains in the building. Traditional lime mortar consists of burnt lime, sand and water and is dangerous (due to the lime spitting as it reacts with the water) and time consuming to make. It was used exclusively until the invention of Portland cement (a vitrified chalk and clay combination) in the early 19th century which remained the principal binder for mortars until the end of the Second World War. The advantage of lime mortar is that it is porous and allows moisture to evaporate, this is very important in conserving old brick or stone. Its disadvantage is that it sets slowly and must not be subjected to frost before it is dry, meaning the building season is limited to March to October. Today a cement, lime and sand mortar is used on smaller buildings and a total cement mortar on many larger sites.

Hand-made Bricks

Bricks were made by hand into the late 19th century. In many cases it was more efficient than investing in machinery and equipment which would be fixed to one site, and then generate transport costs in moving the bricks around the countryside.

Suitable deposits of brick earth are found all over Herefordshire. In the central and east part of the county the underlying geology is red Devonian sandstones and marls, in the north-west it is yellowy brown Silurian mudstones and sandstones. Both bedrocks have a sand, silt and clay content that is very suitable for bricks, being about 25% clay, 50% silt and 25% fine sand, this texture being easy to dig and work and with a clay content that is not too expensive to dry and fire. The problem, however, is to find a sufficient depth, for in many areas the bedrock is near the surface with insufficient depth of subsoil. Alluvial and colluvial (slope wash) deposits and subsoils that overlie the softer bedrocks therefore offered the best sites.

The hollowed out results of clay digging at Burgess Farm, Avenbury

A site also needed to be near a source of fuel and some distance from houses, because of the smoke nuisance. Commons were frequently used, or sites on estate land. For single houses bricks were often made adjacent to the building site.

Permission would have to be obtained from the owner, and in the case of commons usually the lord of the manor, all of whom would require a payment. The 'clay' was dug by an itinerant brickmaker and his labourers in the late summer and early autumn. During the cold weather of the winter it broke down and matured, bricks being made from the following spring onwards and fired in the second autumn.

The clay was smoothed and lumps and larger stones were removed in the mixing process which was carried out by hand or foot in pits or, if available, it was put through a pugmill. This machine had been invented by the Dutch in the 17th century and was powered by a horse. It chopped and stirred the clay, whilst other materials were added at the same time. Additions might be a chalk or other slurry which allowed an unfired brick to dry without cracking; whilst to assist the firing, coal dust and part burnt cinder was added when it was available. Chalk slurry was particularly popular in London in the 18th century, where it also gave the bricks the familiar yellow colour. The idea was to get the clay malleable and workable, but not too wet. In general the brickmaker preferred a wet clay because it is easier to work, and the manager a dry one because it takes less fuel to fire.

In the medieval period there were two ways of shaping the individual bricks, either by laying out a layer of worked clay on sand or straw and then cutting it into appropriate shapes with a spade or similar instrument, or by using a mould. The cutting method was generally used until the 15th century, but thereafter moulding was the main technique used.

Brickmaking lends itself to automation. When it was done by hand in medieval times and with minimal equipment, high rates of production and uniformity were achieved. A team of people would work, paid by output, with each task separated. The brickmaker was the highest paid and a good worker could make over a thousand bricks a day. It is hard work using the upper part of the body and most brickmakers only remained in the trade for about eight years.

The process used a moulding stool, table, two heaps of dry sand, a tub of water, a board called a stock board which in later years had a raised 'kick' or frog (hence the term stock brick, which is merely a standard brick), brick mould and three sets of pallets. A person called a feeder brought the clay from the pugmill, where the clot-moulder, usually female, sprinkled the stool with dry sand, and taking some clay, kneaded it roughly into the shape of a brick. She passed it to the moulder on her left who dashed it with force into

Illustrations from William Pyne's Microcosm *of 1807, in which he gave a 'Picturesque Delineation' of a multitude of subjects in 600 drawings. The three on this page are of brickmakers. Women, men and children, often bare footed, dig clay, push barrows and mould bricks, working under a makeshift shelter. Clay is dug and bricks moulded at the same time and in the same place. Probably making around 2,000 bricks a day, such a team would, in a spring season, make 180,000 bricks for firing in the autumn, clay and fuel allowing, enough for eight small houses*

This second set of scenes by Pyne of brick kilns shows a much bigger operation. Clamp building is shown, it is an all men affair and they are working hard. Horses, carts, ladders, a horse powered pug mill and water pump are all being used. The clamp is high, it needs a ladder to build up and take down. In this vignette you could believe that the team produces a high output of around 3,500 bricks a day. Such a team would be quite capable of making the bricks necessary to build a Hanley Court or Nieuport House

his previously sanded mould. With his strike, a stick moistened with water, he removed the surplus clay throwing it back to the clot maker. He then turned out his raw brick onto a pallet, slid it along to his taking-off boy and re-sanded his mould ready for the next brick. Two people wheeled away the bricks on a hack barrow, and set the bricks on the hack to dry (see Plates 4 to 11, and figures on previous pages).

In Victorian times the moulder and his team would make about 3,500 bricks a day between 6am and 6pm. This is equivalent to about five per minute, or twelve seconds each. Piecework pay gave people the incentive to produce as many bricks as they could.

From about 1690 the practice of making a 'frog' or indentation at the top of the brick appeared in some parts of the country, though it doesn't seem to make an appearance in Herefordshire until late in the 19th century. The origin of the term is obscure, though it seems to relate to the similar frog, or hollow, in a horse's hoof. It was a significant improvement with many advantages. It meant less clay was needed and it forced the clay into the corners making for a sharper brick. Its main advantage though was during building. The frog faces upwards, holds the mortar and speeds the process. Bricks without frogs must be dry or else they 'float', i.e. the mortar won't stick. Having said this, many, if not most buildings in Herefordshire are made of bricks without frogs and bricklayers we have spoken to have said they are fine without frogs. Indeed modern wire cut, extruded bricks have no frog though they have three or more holes which partially fill with cement as one brick is laid on another. Frogless bricks seem to cause no problem with speed of laying or strength of bond.

The raw bricks were stacked to dry in rows eight high in the hack, and protected from rain, frost and heat by straw. When half dry, they were 'scintled', placed in herring-bone fashion, to allow the air to pass freely between them. Normally this would be in fields in long rows, with straw between layers (that occasionally left impressions) and out in the open. The bricks must be dried properly: if not, during firing the water expands in the centre of the brick causing it to fracture or explode. In the height of summer two weeks is sufficient to dry bricks, but at damper times it might take several months.

Bricks would then be fired in clamps or kilns. In the former, when they were quite dry they would be set in vast heaps called clamps containing as many as 100,000 bricks or more, that could stretch across a whole field. Fuel in the form of coal, charcoal and the brickmaker ever frugal, ash, containing a proportion of coal dust, was distributed in layers between the courses of bricks, with especially thick layers at the bottom. The clamps were then encased with previously burnt bricks and covered in clay to hold in the heat. To light the clamp, several flues were left open and filled with faggots. When the clamp was burning well, the flues were stopped and the clamp allowed to burn for a further three to six weeks.

Bricks on the outside of the clamp were under burnt, called burnovers they were set aside for re-burning in the next clamp. Those that were only a little under burnt were called place bricks and were sold off cheaply. Bricks near the live holes that were partly melted were called clinkers and were sold by the cart load for rockeries in gardens. Good partial vitrified bricks were used as damp proof courses. There was much further sorting of the bricks by quality, with as many as ten or a dozen different grades resulting from one firing.

Brick Colour

Brick colour comes largely from the firing, so it is hard to tell the source of brick if relying on colour alone. This can be seen by comparing bricks across the county and the geology. Most are a deep red colour despite the origin of the clay sources (see Plates 1 to 3).

The red colour of brick comes from the iron oxides found naturally in all soils, subsoils and bedrock. Iron constitutes about 5% of the earth's crust. Goethite (FeO(OH)) is the most common iron mineral and gives soil its brown/yellow colour. Hematite (Fe_2O_3) is slightly less common and gives the bright red colour, while larger crystals of hematite give a purpley colour. Both hematite and goethite are present in all English soils and are stable in oxidizing conditions (i.e. where oxygen is present).[1]

If water is driven out of soil and oxygen is present in hot conditions, goethite changes to hematite; this can be at temperatures as low as 45°C. In bricks fired up to 950°C the iron oxide turns to hematite and the bricks are red, the slight variations being to do with the amount of hematite or the presence of other minerals. At very high temperature further chemical changes turn the Ferric iron Fe^{3+} forms of hematite and goethite to ferrous iron Fe^{2+} which has a grey colour. A reducing atmosphere, one where oxygen to the kiln is restricted, produces the same affect. At 1000°C the bricks go brown rather than red while at 1500°C the brick turns to the grey of an engineering brick.

The yellow of 'London' bricks (actually made in North Kent and South Essex), is deliberately achieved by adding chalk slurry to the clay which makes a yellow colour that was preferred at the time. Sand was also sprinkled on green (i.e. unfired) bricks to produce different colours. Clamp fired bricks, particularly those fired with wood, have variable colours that match the variable firing. Today dyes are commonly added to bricks.

The underlying bedrock of Herefordshire consists for the most part of Devonian rocks, with Silurian bedrock in the north-west and in the Woolhope inlier. The yellowish Silurian bedrock makes a slightly yellower brick than the Devonian but it is still a red brick. Alluvial clay makes a yellow, greeny brick presumably because of the different minerals present.

Clamp fired bricks, particularly those fired with wood, have variable colours and shapes due to the different heat experienced during firing and they make for a very pleasing resultant building, though a skilled bricklayer is needed to get the best from them. A good example is Price's almshouses, built between 1630 and 1660. The upper levels and impressive chimneys are made with exceptionally narrow bricks (35mm to 40mm), some improperly fired, showing clearly their clamp fired origin (see Plates 12 and 13).

Kilns were known from the Roman period and were also used in medieval times. Though no medieval brick kiln has been properly excavated in England, they are known from documentary sources and one or two 19th-century excavations. A few examples of excavated medieval pottery kilns also contained brick and tile suggesting these products were sometimes made by potters. Documentary sources suggest some medieval kilns were very large, one in Caister had a capacity of 92,000 bricks, and a medieval yard in Hull had a kiln with a capacity of 40,000. The fuel used was wood or turves.

A typical 19th-century 'intermittent' kiln, so called because it was re-fired with each brick load, is the beehive kiln (see photograph on p.27). This type was very common but eventually became obsolete with the invention of the continuous 'Hoffman' kiln in 1858. A typical beehive kiln took about 7,000 bricks and used about 1.5 tonnes of coal. It had a series of fireplaces set around the wall that were loaded with fuel, sealed and fired using tapers. Experience has shown that a front loading fireplace left open for wood allows a good control of the fire. The ideal was to fire all bricks evenly to 950°C, hold this temperature for 12 hours and then cool it over three days.

Clamp firing is less exact than kiln firing and there is likely to be more wastage with over- or under-fired bricks. Once alight, however, they could be left unattended and so were less time consuming than kilns that required constant stoking. Against this, bricks in clamps were slower to bring to the market than those fired in a kiln. Before going to the clamp, they were up to six weeks air drying and the firing took a further three or more weeks. A kiln-fired brick was fit for delivery one week after it went into the fire.

It is generally assumed that smaller brickworks set up for a particular project would have used clamps but whether to fire in kilns or clamps was largely a matter of choice.

2 Bricks appear in Herefordshire

Today Herefordshire has a comparatively low population, but in the past it was similar to other agricultural regions in England. Like elsewhere, the population declined at the Black Death, slowly recovered over the next four centuries and grew rapidly in the 19th century. What was different in Herefordshire is that the town population mimicked that of the countryside's until the 20th century. It was only then that the county's urban population proportionately expanded. The significance of this as regards brickmaking is that house building and brickmaking took place throughout the county, not just in the main city of Hereford.

The earliest use of bricks in the world was probably in Iraq in what was ancient Mesopotamia in the second millennium BC. These bricks were made of mud, usually tempered with chopped straw and usually sun-dried. Where available they were mortared with bitumen (a natural petroleum product).

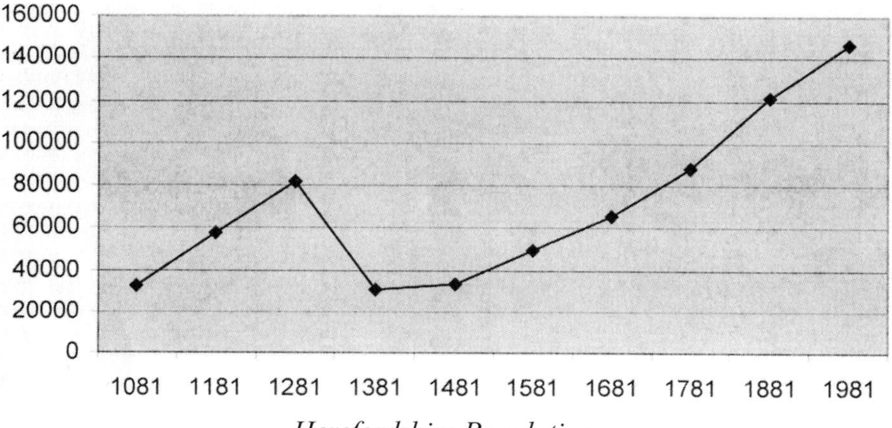

Herefordshire Population
assessed from 1086 Domesday Survey, 1377 Poll Tax;
1512 Subsidy Tax; 1664 Hearth Tax
and 1881, 1901 and 1981 Censuses

Fired bricks were used only occasionally, reserved for decoration on important religious and political buildings.[2]

The earliest bricks in England date to the 1st century AD, in the Roman period. In Herefordshire Roman bricks are rare items though it is quite common to find roof tiles called *tegulae*. *Tegulae* were large (about 40cm x 26cm x 2cm), very heavy (about 6kg) and flat. Curving tiles called *imbrices* covered the joints. Bricks called *bessales* were smaller, about 20cm square and around 4cm thick. Their main function was to create the small pillars that formed part of the underfloor central heating in villas, but they were also used to build arches and as the bonding layers in walls between stone courses. The Romans did not build solid brick walls in this country.

Roman bricks were made in kilns not clamps, which suggests the idea of bricks developed from pottery. About 50 Roman brick and tile kilns have been found in England. They are always square or rectangular and range in size from 1.8m to 5m square. If this type of kiln was 2m high and filled to capacity, taking account of space for fuel, the largest could have produced about 25,000 bricks in a firing, the smallest about 3,000. Most known sites were built into a hill slope, which meant loading was easier working from uphill to down, lifting the raw bricks over the wall, with the stoke hole on the downside. The fuel used was wood, charcoal or straw.

Though there is no clear evidence of the wood coverage in Herefordshire in the Roman period it is likely that there was plenty: enough certainly to make bricks. Tiles from the only known Roman brick and tile kiln in the vicinity, at Leigh Sinton just over the border in Worcestershire, had the paw prints of wolf, fox and stoat. The presence of these animals suggests a landscape with plenty of wood and scrub. The Leigh Sinton kiln measured 3.5m by 2.7m and produced bricks of 260 mm x 260mm x 50mm—a large narrow brick, roof tiles of 400mm x 300 mm x 25mm and imbrices and box flue tiles (used to take hot air for central heating). It could have produced about 4,500 bricks in one firing.[3]

Roman brick has only been found in four villa sites in Herefordshire.[4] This brick may have been made in the county, but given the specialist nature of the craft, it seems more likely that it was imported, either from Leigh Sinton or from Gloucestershire where, although no brick kiln has actually been found, 'Severn Valley' pottery was made and exported to counties in the West Midlands.[5] Hypocaust (pottery pipes used to carry hot air in underfloor heating) and tiles are more frequent finds. Roman tiles can be distinguished from worn tiles of later periods by its softer feel, due to lower firing temperature, and by its larger size and different shapes. Roman tile is still comparatively rare in Herefordshire which suggests that most people continued to live in native wood and thatched round houses rather than romanised villas.

There was no brick produced in the early medieval period in England between AD 410 and 1000. When the Roman military presence, administration and culture left England around AD 410 the art of brickmaking was lost, although it continued on the Continent and in other parts of the world. To us it seems astonishing that such a useful product was not used for over 600 years. The lack of brick is actually one of the most illuminating illustrations of just how different the Roman way of life was to the native British and incoming Anglo-Saxon culture. The Anglo-Saxons often re-used Roman brick, tile and stone as part of the wall fabric of their churches, but they did not make it. In Herefordshire there is actually no evidence of any reused Roman brick or tile, nor of any being made before the 14th century. It has been claimed that Roman brick was used in the 12th-century walls of Putley church, but our own analysis of this is that it is burnt stone and not brick at all.

Medieval brickmaking developed from the tile industry and early brick looks very similar to tile, the difference is only in the depth. Tiles were made in England from around AD 1000 for church floors and roofs.[6] From the 1220s 'flaunderstile' (floor tiles), 'walteghell' (wall tiles) or *'tegulae murales'* (wall tiles) were imported into the east of England from Flanders where this thicker material, brick in all but name, had been used from an early date. The first definite reference to *making* brick in England was in 1335 when 18,000 *tegulae murales* were made for Ely Cathedral at a cost of 12d the thousand. The earliest mention of the word 'breke' is around 1427.

In Herefordshire, houses continued to be built in stone, wood and mud, using one of the two main methods of building houses, frame or mass. In frame buildings walls are made around a cruck or a box frame of wood (or in recent years, steel) and the function of the walls is merely to keep out

Tiles

About 13 tileries are known to have existed in the counties in and surrounding Herefordshire in the medieval period, but only four types turn up in Herefordshire itself.[7] Floor tiles are found in medieval urban deposits, but they are not common and where found there is never enough to make a floor. It seems that tiles were confined to certain places, possibly around the hearth or by the bed, and that the rest of the house had a floor of hard, brushed clay. There were no floor tiles even at the late 15th- or early 16th-century Ford Abbey Farm near Leominster,[8] which was a comparatively high status site with a house and barn, each of three bays. Tiles are found in churches, but even here they were restricted to certain areas and probably didn't cover the whole floor. They often carry coats of arms suggesting they were sponsored by the lord. The impression given is that tiles were expensive and special.

Roof tiles are also rare in medieval deposits of the 12th to 14th century but became more common in the later medieval period, from the late 15th century onwards, but it is ridge, not flat tiles that are found. Ridge tiles are placed along the top of the roof over the thatch, stone or possibly wood. Flat roof tiles, used to cover the whole roof, are extremely rare across the wider region, though they become more frequent in the 17th and 18th centuries.

In Herefordshire the earliest roof tiles are 'knife-cut coxcomb tiles',[9] i.e. they had short, stabbed crests. They were found in reasonable numbers at Hereford Castle where they have been dated to the mid-13th century or later, and fragments occasionally turn up elsewhere. Plainer ridge tiles of the late 13th to 14th century have been found in Hereford and decorated ridge tiles are also occasionally found. Where these have been petrologically analysed they have been found to be made of the same fabric as the contemporary Malvern Chase pottery. This again implies that decorated ridge tiles were made by potters rather than specialist tilers.

Though medieval tiling remains are found more often than brick in excavations, floor and roof *in situ* survivals are almost non-existent. The only tiles in buildings we have traced are at St. Michael's Church at Croft (some *in situ*), Colwall Church (one only now in a wall), Yarpole Church and All Saints Church, Hereford (two only in a wall). In addition there are two outstanding collections, one consisting of about 500 late medieval

Assorted tiles found at Leominster Priory in 2005

Tiles at St. Katherine's Hospital, Ledbury

tiles made in Malvern, which are *in situ* on the floor of the chapel of St. Katherine's Hospital, Ledbury, and the other is a large collection of 15th century tiles also *in situ* used as wall decoration in the chancel of Malvern priory just across the county boundary in Worcestershire.

The paucity of surviving tiles may be due to the successful Victorian Godwin tile firm based at Whitestone (HSM 11863, see below) that supplied tiles to churches all over the country. The availability of these tiles may have encouraged churchwardens responsible for church repair and renovation to replace church floors. Godwin copied the pattern of tiles at Hereford Cathedral, at St. David's and at St. Katherine's in Ledbury and used them in many of the big church restoration programmes of the mid and later 19th century.

Despite the lack of remains, the surnames of people given in the surviving records of the 1379 poll tax show that tiling was common in the county. Of 2,601 surnames 20 are Tiler or Tyler. Their distribution across the county shows there to have been no centre of production. By the time of the Hearth Tax of 1664, only five Tiler or Tyler names are recorded out of a sample of 4,000, but by then surnames did not indicate profession.

Floor tile at Yarpole Church

Floor tile at Colwall Church

A 14th-century timber-framed house in Weobley

the elements. This method was used for both small and grand houses in the past and medieval examples survive well. Notable places to see them are in Pembridge, Weobley and Eardisley where the 'black and white' box-framed houses are famous, and date to the 15th century. Where brick can be seen in these buildings it is a later addition, replacing the original wattle, daub and plaster.

In mass buildings the walls themselves form the frame. This method of building was used for the many castles and churches built by the Normans, and occasionally for housing. Medieval examples of mass frame houses are rare, larger ones can be seen occasionally but the smaller examples have not survived.

Most people in medieval Herefordshire lived in box-framed houses with wattle and daub walls, that sometimes were erected on courses of stone to protect them from the damp. The house sites were known as tofts, and were set in small plots called crofts, positioned close to the manor court and church, often aligned along a street. Flimsy and made of organic materials, these houses have disappeared without trace and we know very little about them

The oldest Herefordshire houses that survive date to the 15th century. They survive because they were quality houses, built of green, straight, heartwood of oak, cut parallel to the grain. Brick was not used in these houses.

Brass Knoll Farmhouse: a 16th-century mass-framed buttressed house in the Golden Valley

Despite 25 years of archaeological observation of medieval deposits, covering hundreds of urban and rural sites across the county medieval brick has only been recorded on three sites in Herefordshire. One was in 14th-/15th-century deposits found in 1975 from Wigmore Abbey. The bricks measured about 50mm by 120mm and had been made in a sanded mould with

Fig Pikes Farm, the only small medieval mass constructed house known to the authors in the entire county

An example of crumbling wattle and daub panels in a box-framed house

a trimmed upper surface. Another was at Breinton during a 1920s excavation of the defended site adjacent to Breinton church (HSM 928). A third is a recent discovery from a 2004 excavation by Archenfield Archaeology at Mill Street in Leominster.[10] Here brick varying in width between 37mm to 60mm depth was recovered in 13th- and 14th-century deposits. Out of county at Monmouth three bricks were found in a tile producing kiln dated to the 15th century, where the one intact brick was remarkably similar in size to a modern brick—225mm by 112mm by 63mm, but so little was recovered the excavators interpreted it as kiln furniture.

More medieval brick remains may have been found, but brick and tile has normally been discarded in all but very recent excavations. Nevertheless the combined evidence suggests that brickmaking was rare in Herefordshire throughout the medieval period. Even as late as 1615 brickworkers had to be brought into Monmouth from outside the area because there was no local knowledge.[11]

Brick in medieval England was essentially a luxury product—it was different, colourful and because of the amount of fuel it used, expensive. Between 1400–1450 several prestige large brick houses were constructed in eastern England, built to show off wealth, status and good taste to the neighbours. The 15th-century brick Hellens House in Herefordshire (discussed below) was of this type and is an exceptionally early brick house for the West Midlands. This red mansion is startlingly attractive even today when brick is very common. It must have been doubly so when first built.

Brick was used more universally from the 16th century onwards when houses developed chimneys. On standing building evidence, it seems that in Herefordshire chimneys were generally of stone, but with the narrower upper levels often built of brick which is more resistant to damage by heat. Despite its suitability we have found few examples of brick chimneys of the 17th and 18th centuries. This must be because the tall, flimsy construction of chimneys and the softer nature of early brick, means the original has long since been replaced. When the chimneys were recently rebuilt on Price's almshouses (discussed below) in Whitecross Road for example, the original bricks of 1665 crumbled away to nothing.[12] Nevertheless the demand for chimney bricks must have led to the development of a brickmaking craft in the county.

Aside from chimneys, few houses in Herefordshire were made of brick before the 18th century. Looking for 'brick' on the listed building database reveals 1,757 buildings with brick in Herefordshire dating to the 17th century, but the majority of these are timber-framed with brick infill. Studies of individual buildings shows this infill represents later repairs or improvement.

Medieval Transport

Transport was expensive, so building materials generally were procured as close to the building site as possible. This has led to regional diversity in England, where locally produced bricks gave their own character to each county.

The capacity of a larger horse drawn cart was about 800 bricks weighing about two and a half tonnes (bricks were slightly heavier in the past). As it takes over 20,000 bricks to build a small house, it can be seen that making them as close to the building site as was feasible was the most sensible course of action. Having said that, wealthy people in the past did transport heavy items like wood and building stone. The evidence suggests that the small amount of brick found in Roman and medieval archaeology layers was made out of county and transported in.

Fossilised roads, i.e. green lanes, in Herefordshire show that many were very narrow, banked and hedged. Becoming deep in mud in winter in the days before tarmaccadam, most bricks would have been transported by packhorse rather than cart.

A good example of a traditional muddy Herefordshire track

Only a handful of 17th-century buildings survive that were originally built wholly in brick. All are large: Bernithan Court, Llangarron; Heath House, Leintwardine; Hanley Court, Kingstone; Castle House, Eardisley and Moorend Court, Mathon are good examples of substantial 17th-century houses built primarily in brick.

About 50 buildings survive that were originally built partly in brick, including Gatley Park, Holme Lacy House, Fownhope Court and Home Farmhouse, Allensmore. About one fifth of the listings refer to brick chimney stacks, for example Rydicar at St. Weonards, Old Lands at Mathon and Old Court Cottage at Hampton Bishop. The chimneys there today could be original but most have probably been rebuilt.

The use of brick increased in housing in the 18th century, with over 800 listed brick buildings scattered throughout the county, in town and country alike. Castle Street and much of St. Owens Street, Hereford is 18th-century brick, several houses in Broad Street, Bromyard, several in the Homend and Southend, Ledbury and several in Broad Street and High Street, Ross-on-Wye. In Leominster, 18th-century brick buildings lie in many of the main streets, interspersed with later 18th- and early 19th-century buildings. Only Kington is different, here there are more 18th-century buildings of stucco (sand and cement patterned to give the appearance of stone) or rubble, with only a few houses of painted brick. The stone tradition continued in Kington for longer, probably due to the Penrhos quarry that even in 1858 was producing building stone at 1s 6d per ton making it a good deal cheaper than the equivalent in bricks. Typical examples of country 18th-century brick buildings are Hollywood House, Abbey Dore; the hop kilns at Whitehouse Farmhouse, Bishops Frome; The Red Lion Hotel in Bredwardine and a row of small cottages in Wilton Lane, Bridstow.

The 18th century was an interesting time for brickmaking as it was for skills and crafts generally. Pre-canal and railway, industry and agriculture was decentralised. Brick buildings are everywhere, scattered throughout the county, there is no obvious bias towards forest or other regions, or between town and country. It suggests that brickmakers went to the building rather than brick being transported from centres of production. Clay would have been dug on site, coal brought up the river from the Forest of Dean for use in Hereford, and brushwood or coal brought in by pack horse elsewhere. Many Herefordshire tracks bear witness to this packhorse trade—small and sunken, they are too narrow to take a cart and could only be negotiated by packhorse. There is some documentary evidence for the packhorse trade. For example, papers referring to wood sales were measured in the horse load in 16th-century Bringewood. As late as the mid 19th century iron ore was brought to a forge on the Arrow at Pembridge from Wales by mule, and to Downton Gorge from Clee Hill by packhorse.[13]

Extensive research by David Whitehead has revealed about 70 references to brickmakers and sellers working in and around Herefordshire between the mid 16th to early 19th century. Some refer to court cases, others to work carried out. These are discussed in greater detail in Chapter 4.

Brickmaking, Fuel and the Countryside

Brick used a lot of fuel, the equivalent of about five tonnes of coal to make one Victorian house. Herefordshire has no coal and before cheap transport, brick and tile makers burnt 'under' wood, i.e. coppice, brushwood, charcoal or gorse. A well-managed coppice produces around 2.5 tonnes of dry wood per hectare per year. Fresh cut oak has a density of broadly 1.1cm^3/gram. The equivalent calorific value of dry wood to coal is about 1.4 times the weight, and a much greater volume as wood is bulkier. One beehive kiln making 7,000 bricks uses about 1.5 tonnes of coal, which is the equivalent in dry wood of about 2 tonnes. A Victorian terrace house using around 22,000 bricks would therefore need about 6 ha of well-managed coppice, plus the labour to turn this into faggots or charcoal, and take it to the brickmaking site.

Up until the 19th century woods generally belonged to the lord of the manor. They produced timber for building and the coppice was cut on a seven to 20 year cycle. In addition the villagers often had the right to collect fallen wood and graze their pigs. Timber from 15th-century houses proves that there was plenty of good straight oak wood cut from close growing stands of around

Forests and Brickmaking

A forest connection with brickmakers has been noted in Arden and at Feckenham near Redditch in the 16th and 17th centuries, the association being due to the presence of fuel, spare land, clay and the less rigid agricultural regime of the forest. Brickmaking became a secondary occupation of the small farmers who lived there.[14] The same type of small-scale brickmaking also occurred in Malvern Chase on the Herefordshire border. This is known because Malvern brick can be identified due to the presence of distinctive minerals found in the pre-Cambrian rock. It was found in the 13th-century layers of the Manor House excavated at Breinton (HSM 928) for example, which is more than 20 miles away from where it was made.[15] Tiles were recovered from the Castle Moat site in Leominster, that came from Hanley Castle in Malvern Chase, also 20 miles away.[16] The Deerfold/Lingen region that was Darvold Forest in the medieval period, is another known pottery producing area that also made floor and roof tiles. Tile from here was transported as far as the Castle Moat house site in Leominster, ten miles away. The distance these heavy products travelled implies there was no brickmaking in central Herefordshire at this time, presumably due to lack of skill, wood or landlord or commoners' restrictions.

a hundred years old, proving that some woods grew as plantations in this post Black Death (1348), low population period. Pre-Black Death court rolls show that Herefordshire was not an especially wooded county. Population was high for a largely self-sufficient society and there was more demand for wheat than wood, so most land was given over to this. Later, in the post medieval period, it is likely that managed wood expanded in upland areas or poor land, and this was used for industry, including small-scale pottery and brickmaking, but its supply was still limited.

Some brick and pottery making sites around Upton Bishop and Garway in the south of the county were fired with wood and located near coppices. The excavators interpreted the kilns as short lived, perhaps using one season's coppice wood and moving on elsewhere.[17]

The nearest sources of coal for Herefordshire are the Forest of Dean in the south, Wales to the south-west and the Midlands to the east. Coal came from the Forest of Dean by barge up the river Wye, but not without difficulty—the river is notoriously difficult for navigation. Some improvements were made to navigation in the third quarter of the 17th century and coal was sold in Hereford at 11s ton in 1740, rising to £1 6s in 1803.[18] The canal from Mamble near the Wyre Forest to Leominster meant coal from the Midlands also became cheaper from 1796, and brickmaking became economic near the canal and the Wye. As the prices section below shows, bricks were being sold for 17s a thousand in 1728 and one ton of coal could make over four thousand bricks. Outside of Leominster and Hereford however transporting coal in any great quantities remained uneconomic.

Due to the scarcity of fuel people collected gorse. A Sarnesfield account book records that on 27 January 1735 the estate paid £1 9s 4d for 2,200 gorse faggots (16d/100 faggots). The same month £15 18s was paid to a brickmaker and £3 5s 9d to a lime kiln.[19] It implies gorse was collected as fuel in the 18th century, supplementing the fuel supply from coppice.

Medieval Prices and Wages

In 1283, 101,350 Flemish bricks were bought for the London town wall for £22 3s 5d, which equates to 19 bricks for 1d, not including transport. Bricks described as 'walltiles' were being sold for as little as 2s 3d per 1,000 (37 bricks for 1d) to as much as 5s 2½d (16 bricks for 1d) in East Yorkshire between 1303 and 1410, the latter amount including carriage. In Ely in 1335 wall tiles were made for only 12d per 1,000 (83 bricks for 1d) though elsewhere 1,000 tiles cost about 48d (21 bricks for 1d).[20]

Quarried stone at this time was about 2d a cubic foot—an area the equivalent of about 23 medieval size bricks. On these rates medieval brick compared well with the cost of stone, being broadly similar in price if not cheaper. Despite this, bricks were rarely made in Herefordshire or elsewhere in these centuries presumably because people built in the tradition they knew, in stone, wood and mud, and because of the problems of getting fuel and skilled labour.

In the 15th and 16th centuries the price remained remarkably similar at 4s a thousand (21 bricks per 1d), but those made for King's College, Cambridge, which must have been rather special, were being charged at 13s (6 bricks per 1d). In the 17th century with the general rise in prices, bricks were in the mid-teens of shillings per 1,000, and by the 18th century the £1 barrier was breached.

Stone cost 1s 6d a load in central Herefordshire in 1740, including transport. Clay carried the same distance was 1s the load, tiles 1s the load and poles 2s 6d the load.[21] This would suggest that brick had become more expensive than stone in the 18th century, as the clay would have to be fashioned and burnt which would more than double the cost.

The wages for the workers who were unskilled, except for the brickmaker, are often not recorded. Manorial accounts give agricultural rates of pay until around the mid 15th century and some building accounts survive. From these it is known that in the 13th century workmen were paid about 1½d a day and skilled craftsmen about 2½d a day. At this time wheat, the basic foodstuff, was around 352d a tonne or ½d for 4,000 calories, roughly the amount a workman would need in one day. In theory one third of income went on food alone and if s/he had a family this would amount to over half of their wages.[22] However, it is important to remember that prices were different in the past. Wheat was highly priced right up to the repeal of the Corn Laws in 1846, while other items based on labour were very cheap. Until the industrial period, people provided for many of their needs without money through a system of barter.

The comparative cost of raw materials, including transport, in 1740

A brickmaker was probably a smallholder as well, and so would grow much of his family's food.

Historians have built cost of living index tables based on the price of grain, fuel and tallow (used for light), to work out how well off people were.[23] By taking a ten-year average a pattern emerges. Wages rose for the agricultural worker from around a rate of 1d a day, to 2½d in the mid 14th century, 3½d in the 15th century, 8d in the 16th, 12d in the 17th and 20d in the 18th century. However the buying power of wages didn't rise until the 19th century because prices rose as well. In the 17th century wages are estimated at only 75% of their purchasing level in the years 1260–1349, the period of highest population before the late 18th or early 19th century.

Records from 1651 in Chelmsford, where the daily wages of 'Makers of Brick and Tile and Burners of Wood Ashes and Lime' were fixed, show the year was split. Between mid-September to mid-March the clay was dug and allowed to mature. Between mid-March to mid-September the clay was made into bricks, fired and sold. In the first period the wages were 6d with meat and drink and 12d without, in the second period the wages were 8d per day with meat and drink and 16d without. These men could employ servants and labourers at 5d and 10d and 6d and 11d respectively. Interestingly the ratio between wages and the cost of food appears to be much the same as in the medieval period, wages being about double the price of food for the day.

Daily wages were later replaced by piecework. In 1683 away from London, a good moulder, working alone could produce 1,000 bricks on a summer's day of 14 or 15 hours. With a man to temper the earth and a boy to carry the 'hacks' he could turn out 2,000 or even 3,000 a day. He was paid 4s per thousand of which he retained half. Fifty years later in 1723, the moulder was being paid 6d a thousand, the 'bearer off' 4d, 'he that tempers the earth' 4d and 'he that digs it' 6d, and 9,000 bricks was considered a good day's work.[24]

If these figures are correct and normal it shows a reduction in the quality of life of the brickmaker in the later 17th century. He now had to work twice as hard for the same money. He may well have had a reduced status as well, compared to the medieval period when brickmaking skill was a rare commodity and brick a luxury item.

3 BRICK TAKES OVER

The steady population rise from the 17th century—in Herefordshire from 65,000 people to about 110,000 in the mid 19th—led to a demand for houses that could not be satisfied by traditional methods of building. During this time many brickmaking machines were invented, versions of which are still in use today. In the 19th century brick was cheaper than freestone but not cheaper than rubble if the quarry was near to hand. More to the point is the comparison with timber. Cruck frame construction had all but ceased by the 17th century in Herefordshire due probably to loss of the park type of landscape that produced spreading oaks from which crucks could be cut. Box-framed houses using oak taken from managed woods continued into the 19th century and then all but ceased. This was due to a combination of insufficient timber to accommodate the population increase, the expense of timber compared to brick and change in fashion.

These factors meant that houses were increasingly built in brick in the 18th and 19th centuries, and from the Victorian period onwards almost exclusively so.

The first brickmaking machines imitated the hand moulding process, but speeded up the process. Later machines extruded the clay from a die in a measured length which was then cut by wire to brick size. By the 1880s steam was being used, and machines could produce 1,000 bricks an hour. However the cost of the machinery was high, and for this reason smaller brickworks, the most numerous type, were slow to invest. Powered pugmills were considered a necessity by the end of the century, but in general, labour was so cheap that hand production continued throughout the countryside.

As has been noted earlier, machine mixed bricks can use a higher percentage of clay than those mixed by hand and the finished brick is potentially of better quality being harder and denser. However, machinery doesn't necessarily imply high clay content, the manufacturer could put all manner of things in a brick, just so long as there was sufficient clay to hold it together. The authors have a brick retrieved from the Blaenavon area, for example, that is made from quarry rubble.

By the 1860s, the Victorians liked the precise finish of machine-made bricks with their uniformity of colour and shape, and encouraged bricklayers to work with them and use thin, straight joints. These mortar joints could be sufficiently threadlike to have no toning down effect on the impact of brick colour. A good example of this work can be found in John Street in Hereford.

Taste inevitably changed and people again wanted the appearance of hand-made bricks but with the precision of machine-made. Attempts were made to use soft mud in machinery, but it was not until the last year of the 19th century that a machine, patented by R.A. Norris, was able to do this. So good was this machine that it was in full production for the next 35 years.

The wire cut process was first used in the 1840s, but the first fully operational machine was designed by Richard Bennett in 1879, and used for the following 50 years. The wire cutting process soon spread from England all over the world. The first designs of the wire-cutting machine pressed a quantity of clay through the die by a piston, called a 'stupid'. At the end of its stroke it needed to be withdrawn and refilled with clay, so it was not continuous; this may have been the source of its nickname. In the next development, a screw or augur was turned, the raw clay being inserted continuously at one end and the brick plug coming out of the die at the other, to be wire cut, a process that has continued to this day. The resulting brick can be distinguished from pressed bricks by the three or more holes left from the augur. The movement of the clay to the brickmaking machine and the removal of the formed bricks was finally completely mechanised soon after the First World War.

An example of a German extruded brick

Not all brickworks adopted total mechanization. Old fashioned team work still continued and produced high outputs. In 1923, at Broad & Co. Ltd. of West Drayton, a team of brickmakers, called a stool, comprised the moulder, the offbeater, the temperer, the flatty, the pusher-out (boy) and the barrow loader (boy). An ordinary team

A Hoffman Continuous Kiln on the outskirts of Birmingham (Birmingham City Archives)

could produce about 700 bricks an hour, equivalent to 6,300 a day of 9 hours and 35,000 a week of 50 hours, rising to 47,000 for an exceptional stool.[25]

Another 19th-century development was that of the Scotch Kiln. This was a rectangular brick-built kiln with fire holes along the sides and open at the top, and with doorways in each end. When full the top was covered, the flues and entrances were stopped and the kiln fired. There are no longer any traces of these in Herefordshire but they were probably used in all the bigger brickworks.

Further improvements were achieved with the Hoffman Kiln. This allowed a down draught of heated gases which gave stricter control and produced bricks of consistent quality. It also used less fuel, in that the round kiln allows the fire to move from chamber to chamber, heating some and allowing others to cool, thus becoming a continuous process. Green, i.e. unfired, bricks were placed in the cool chamber, to be dried properly as the heat front moved, then they were fired and finally cooled for removal. This idea of large continuous kilns is the method used today, although gas, oil and electricity are the fuels used rather than coal.

The Leominster canal, bringing coal from the West Midlands, opened in 1796. Tramways bringing coal from Wales were opened to Hay in 1816 and to Hereford in 1829. These were short-lived affairs, however, superseded by the railways which were built from the 1850s and whose tentacles stretched

A brick clamp being formed at Acton Scott Working Farm in June 2004. The top picture shows the clamp being constructed with a fill of coal, with holes for igniting the fuel and the subsequent drawing in of oxygen left around the base. The bottom picture shows the clamp ready for firing

further into the country. Brickworks could now be located near stations, to where fuel could be cheaply transported, and bricks taken away. Large brickworks thus developed in Hereford, Ledbury, Bromyard and Leominster. Housing of the period 1860 to 1935 was largely made of bricks, but most of these were manufactured in these larger factories.

Ultimately cheap freight costs led to the demise of Herefordshire brickworks altogether, as it eventually became cheaper to import Fletton bricks from the east of England. In the late 1880s the clay at Fletton near Peterborough was found to be shale-oil bearing and in consequence needed less fuel for firing. Indeed, with its 10% fuel content the brickmakers found it almost burnt of its own accord. The Fletton brickmakers cut down production costs by maximising on their advantage and investing in machinery and buildings. Geologically the material was Oxford Clay and extended in a deep stratum some feet below the surface from Yorkshire to Dorset. The discovery was the beginning of the end for most other brickmakers.

The depth of the Oxford Clay is very great, 60 metres being quite common. The depth required operating costs of a large scale, which in turn meant large sales. Instead of producing 2.5 million bricks over twenty years as in a typical Herefordshire brickworks such as Colwall, sales were in the tens and hundreds

A Beehive Kiln at Acton Scott Working Farm

of millions a year. This was a huge output, but the demand was there. For example, 25 million bricks were used in the construction of the War Office in Whitehall between 1903 and 1904.

The competition from London bricks, or the Fletton, put smaller brickmakers out of business. In Herefordshire the Colwall Brick and Tile Works were typical. It existed from 1880 to 1900, and only stopped trading because fuel bills became too high. Most of the remaining Herefordshire brickworks went out of business in the years up to the Second World War.

Fletton bricks had a tremendous impact on our landscape. The ability to build with cheap clay bricks enabled Britain to become a land of brick-built waterproof houses, and not one built of hollow bricks and rendered for waterproofing as is the case in some countries on the continent.

'London' bricks were used almost exclusively in Herefordshire after the Second World War. A constant supply arrived by train at all three stations of Hereford (Barrs Court, Moorfields and Barton) where they were unloaded by the factors Sweetman and Sons or Bristows who then sold them on to the builders.

The brick industry today has not yet succumbed totally to globalisation. Brick is still largely produced in England, though the business is concentrated in the hands of just a few companies. In 1946 it was estimated there were 1,350 brickworks employing 40,000 people, in 2005 there were just 103. In the 1960s, 50% of the brick was Fletton type, which reached a peak in the

Often houses were built in the place from where clay had been dug, as with the house pictured here in New Street, Ledbury, about 1.2m below surrounding land

1970s when 3,500 million Flettons were made each year. In more recent years other brickworks have overtaken the Fletton market which has fallen to 25% of the total.

In the late 20th century the English output has been broadly:

1950	8 billion per year
1974	6 billion per year
1988	4.7 billion per year
2004	2.8 billion per year[26]

The fall in brick production is because modern houses use fewer bricks in their construction. Most housing today is merely faced in brick, the internal walls being of block or timber frame. A small to medium sized house will now use about 9,000 bricks as compared to around 22,000 in the 19th century. Typically a new housing development such as the Bradbury Estate, Hereford used 4.5 million bricks for the 500 houses that were built in 2005/6, The same number of houses in Victorian times would have used 12.5 million bricks.

Modern wall construction showing the single brick outer skin, with light-weight blocks for the inner skin (now also often built as a timber frame)

In the 1930s the average cost of building a house was between £334 and £409, brick comprising about £40 to £50, or 12% of the total. In 1988 the cost of bricks had dropped to 3 to 4%. In 2005 bricks cost about £220 per 1,000 delivered to site, plus VAT. The price of a three bedroom new house in Herefordshire using about 10,000 bricks is around £190,000 with the cost of materials being about £60,000, bricks representing about 1% of the cost of the house and 4% of the total materials.[27] So brick cost has fallen due to technology, economy of scale and cheap transport but primarily because less brick is used in each house.

Brickworks today run 24 hours, seven days a week. Despite the machinery however, the observer can see the same basic processes as occurred in medieval times. Clay is dug from geological strata and, for example at Baggeridge Bricks, Hartlebury, is comparatively high in clay—as much as 40%. It is moved by conveyor belt from ground to a holding area, to the mixer, to the extruder or moulder, on to dryer, the oven, the packer and thence to the lorry. Throughout the process they are not touched by human hand. Baggeridge Bricks currently makes about one million bricks a week or 16,000 bricks a day per one operator, which represents about 9% of the UK market.

4 Brickworkers

Even where machinery was introduced, in the small rural brickworks brickmaking remained a seasonal occupation with no work in the winter. The cold and wet weather made it difficult for work and was used as the time when the clay matured, so traditionally all the workers were laid off. This practice remained well into the 19th century; insecurity was typical.

Wage rates and conditions of brick workers are not easy to find. Although brick invoices survive, the money is paid to the proprietor and the number of people employed to produce the bricks or the hours they worked is not given. It is known that in the Victorian period the difference between skilled worker and labourer was jealously guarded, though in truth there was not always that much difference in the amount of money actually received due to deductions exacted by employers for various items of the trade. By this time, brickmaking was no longer considered a specialist trade and brickmakers' wages are likely to have been similar to those of day labourers. Skilled workers were paid between 15s and £3 a week. A skilled mason was paid 15s a week, and this rate remained the same for about 100 years in the 18th and 19th centuries. The 1818 *Book of English Trades* states smiths earned 30s a week, carriage makers £1 to £2 and body-makers (of carriages) £3 week. Unskilled workers were commonly paid by the day indicating the even greater level of insecurity of employment than the skilled worker's weekly pay. In 1843 a quarry labourer was paid as little as 6d a day in Westhope quarries, while Hereford builders were paid 1s to 2s 6d a day. This was more than the 7d a day paid to Herefordshire women agricultural workers and slightly better than the 1s 8d a day average in England for an agricultural labourer.[28]

Work for the unskilled in Victorian England implied long hours, low security and subservient status. Brickmaking was no exception. There are several examples in the next chapter of Herefordshire brickmakers who switched from master to man to farm labourer. The work was hard and outside, although people working in the smaller businesses were often self-employed which gave some measure of independence. The machines made the bricks, but did nothing for the processes between the pugmill and the machine, nor the removal of the formed bricks to the drying areas.

Even in Edwardian times teams of men were working physically hard for ten hours a day. Most brickworkers were poor, Charles Booth found in his survey of London workers in the late 19th century that 55% of bricklayers lived in poverty. Dickens describes two brickmaking families in *Bleak House* (1852–1853) who are forced to move around the country with their families and live in very bad conditions (though this was partly due to the drunkenness of their husbands). Work conditions were hard for many people of course, and brick workers were not alone. It took many Parliamentary Acts to reduce the working week to tolerable levels. In 1834 the Ten-Hour Day Act was passed, in the 1870s this was reduced to a nine hour, six-day week, and in 1920 to an 8-hour 5½-day week.

The one and only photograph we have found of brickmakers at Much Cowarne brickworks in Herefordshire shows a group of fit men and boys working in a field in the late 19th or early 20th century. They are in warm clothes and though the boys have no shoes, they are reasonably well fed. In this particular photograph there are no women, but census records show there were plenty of women brickmakers, even though this actually went against work regulations. The 1864 Factory Act that made it unlawful to employ women or children under ten, exempted brickworks, but in 1871 the Act was extended to include brick and tile yards. Up to age 16 boys could work for three days a week but not before 6am nor after 9pm, and an extra half hour must be given for a meal at 5pm. (Bromyard & District History Society Archive)

Not all was bad news, however. Labourers working in the Hampton Park Brickworks in Hereford in the late 19th century lived in pretty, though very small cottages on the edge of town a stone's throw from their place of work.

Wages and conditions improved for working people throughout the late 19th and 20th centuries, though by as late as 1946 the situation for brick-makers was poor. Only £4 7s was paid for a 60-hour week,[29] 1s 5½d an hour, this drew many foreign workers, since local people were reluctant to work at this rate. The agricultural worker was on a similar wage at £4 14s for a 47-hour week, though some farm workers would have had indirect payments of food, making them slightly better off.

Today brickmaking is largely automated and people work under cover, operating machines and/or computers. The modern brickmaking plant now works round the clock as equipment is too expensive to be shut down.

Brickworks could not function without people, but finding information about them is not easy. Just as we have been unable to find more than a handful of pictures of Herefordshire brickworks because they were considered commonplace at the time, so too were the workers mainly ignored by those who could draw or who had a camera, and as a consequence their lives remain unknown. As with all research, the further back you go, the sparser are the records. Nevertheless, we have been fortunate in tracing some Herefordshire people who were brickmakers.

Early Herefordshire Brickworkers

Grafton was an area where there was a long lived brickmaking tradition. **William Fletcher** was a brickmaker at Grafton in 1835 when the two works recorded in the *Gazetteer* were in production. (The Grafton Inn was formerly known as The Brick Kiln Inn).

There are six known brickmaking sites in **Leominster**, indicating a long established trade. **John Rogers** was a brickmaker there in 1780, and **Robert Evans** was another who supplied 3,000 bricks for Arrow Bridge in 1820. He had earlier been in partnership with a **Mr. Yeats** around 1799. A name that appears many times in this book, especially in the *Gazetteer*, is **Tunks**. In the 1780 Poll Book, Richard Tunks is recorded as a brickmaker from Leominster.

There is only one possible brickmaking site in **Luston**, a croft that was called Brickton Close in 1840. In 1783 however **Benjamin Beach**, a brickmaker of Luston was declared bankrupt, so perhaps this was his house and with his bankruptcy, brickmaking ceased. **Sarnesfield** had been a brickmaking site for many years before the tithe maps were drawn in the 1840s. **Samuel Beach** of Sarnesfield was paid £15 18s in 1736 for bricks.

Records of brickmakers often derive from estate records as in the following five examples of Moccas, Shobdon, Tyberton, Hellens and

Stoke Edith. **Thomas Williams** was making bricks at Sarnesfield for **Moccas** in 1736, for which he received £36 9s 6d. Another **Thomas Williams**, possibly his son, was making bricks at **Moccas Court** when the present building was going up in 1776. There was more than one brickmaker at the time of the rebuilding, as **Mr. Hooel** (possibly Howell) was a brickmaker there in the same year. **Will Mathews** (*sic*) was paid £40 12s 11d. for bricks at **Shobdon Court** in 1721 and the following year **Humphrey Perks** was making bricks at **Shobdon**. There is a direct reference to **Tyberton Church** when **Mr. Lawrence** was described as burning bricks for the chancel and in 1791 (see chapter 6), **James Williams** was doing the same at **Hellens House** (chapter 6). These all refer to specific events in the building of these three houses, and it can be reasonably claimed that the two Williams and Lawrence were the master brickmakers in charge of the works.

The front elevation of Moccas Court, made of brick in 1776

Stoke Edith House also has a very long history of employing brickmakers, as the owners frequently made alterations and additions to their house. The earliest record we have is of **Peter Simon**, brickmaker, who was there between 1696 and 1698. For some speciality work, **E Cook**, brickmaker, supplied arch bricks in 1793. **John Herring** who is referred to below under the Cathedral and the County Gaol, was supplying brick and tile to Stoke Edith between 1793 and 1800 and in the same years another brickmaker, **Robert Jones** was working there. More speciality work was prepared by **John Fletcher** when he supplied dress bricks for the pillars of the Hereford Lodge 1796.

Hereford Cathedral has a long history of employing brickmakers. The earliest record shows **Mrs. Cooke** supplying brick under the Cathedral Fabrick Account 1720–21. In 1728 **Mr. Wood** provided lime and 2,500 bricks at 17s per 1,000. Another **Cook** supplied bricks under the Fabrick Account and received £5 9s in 1780–81, £27 14s in 1782–83 and for a wall in the accounts of 1792–93 a further £2 8s. In 1787 he was supplying bricks for the Cathedral and the Cathedral Library, then in 1805 he was supplying

The Bishop's Palace, Hereford, faced in brick in the 18th century

more bricks for **Ethelbert's Hospital**. A final note for Mr. Cook states that when he died at Kings Caple he had brick and tiles among his goods.

John Herring was a brick- and tilemaker from Tupsley, who supplied **St. Ethelbert's** in 1805. This was a year of much work at the Cathedral as we have records of **Mr. Perry** supplying bricks, **Mr. Wainwright**, who was described as a tinker, a merchant and a brickmaker supplying yet more bricks and **Dr. Cam** providing his quota. We think this abbreviation is unlikely to mean a doctor, more probable it is a shortening of his first name.

John Freeman of Stretton Grandison was thought to be a brickmaker. He certainly supplied 500,000 bricks at **Ashperton/Canon Frome** in 1797 at 20s per thousand. Such a large quantity of bricks suggest that he was supplying the **Hereford to Gloucester Canal,** especially since the 400m (¼ mile) Ashperton Tunnel is here. In Hereford itself, **John Herring**, we assume the same as above, supplied bricks for the **County Gaol** between 1796–98, and **Isaac Lee** was a another who supplied bricks for the **Shire Hall** in 1815.[30]

Later Herefordshire Brickworkers[31]

The brickworks that follow are referred to by their HSM number or a reference to a brickyard in a settlement. You will find all the brickworks we have traced listed alphabetically by parish in the *Gazetteer* at the end of this book.

Almeley: James and Elizabeth East, and their son James, all described as brick burners living at Meer (Mere Common).

Blakemere: There are three sites in the gazetteer, Blakemere Brickworks (HSM 31572), Kiteswood Brickworks (HSM 31573) and Brick Meadow (HSM 32295), but only one brickmaker, Thomas Griffiths who lived at Raven.

Brick Sizes[32]

Bricks are of a size that can be easily carried and handled, and so have remained remarkably similar for thousands of years. The exception was the larger bricks made in the Roman period. Some of the bricks at this time, often called tiles (*tegulae*), were flat and large but there were a number of standard sizes which were based on the Roman foot, and ranged from about 6 inches square to 2 feet square. This appeared to set the standard when brickmaking came back in the early Middle Ages (see the list below). However portability soon became a factor, and the next in the list is of a more practical, and also a more aesthetic shape when seen in a building.

From 1571 brick size was regulated by government, the sizes changing slightly by order, right up to the present day. Despite the regulations, out of 50+ bricks from different brickyards we have measured, none are the same size. Either people ignored the regulations to some degree, or the making process lead to shrinkage and distortion.

Early brick is narrow, 50 mm or less while brick from the early 18th century onwards is thicker, 70mm or more. The brick tax (see the box on p.38) did not seem to affect the size much in Herefordshire, brick was already of a 70mm or so depth before its introduction in 1784. Occasionally particularly deep bricks were made, a chapel on Ridge Hill, Hereford for example or the extension on the small house beside Little Dewchurch church, but these are unusual and post date the brick tax in any case.

The British Standard of 1965 was 8⅝ x 4⅛ x 2⅝ inches, metricated in 1969 to 215 x 102.5 x 65mm. Sizes are similar to the modern standard through time. Here are some examples from the 13th to the late 18th century: inches (millimetres)

1200	Little Coggeshall Abbey	12 x 5½ x ¾ (301 x 140 x 45)
1280	Allington Castle	9 x 4¼ x 2 (228 x 107 x 51)
1380	Thornton Abbey, Lincs	11 x 5½ x 2 (279 x 140 x 51)
1480	Bishop's Palace, Hatfield	9 x 4¼ x 2 (228 x 107 x 51)
1520	West Stow hall	9¼ x 4 x 2⅛ (234 x 101 x 53)
1568	Statute brick	9 x 4½ x 2½ (229 x 115 x 63)
1571	Statute brick	9 x 4½ x 2 (229 x 115 x 51)
1625	Statute brick	9 x 4 x 2¼ (229 x 102 x 57)
1631	Kew Palace	8½ x 4¼ x 2¼ (216 x 107 x 57)
1660	Price's almshouses, H'ford	9½ x ? x 1½ (240 x ? x 35)
1635	Raynham Hall, Norfolk	9¼ x 10 x 4¾ (235 x 254 x 120)
1705	Queen Anne's Gate, London	8½ x 4 x 2¼ (216 x 101 x 57)
1725	Place bricks, general	9 x 4½ x 2½ (229 x 114 x 64)
1725	Stock bricks, general	9 x 4½ x 2 (229 x 114 x 51)
1760	Stock bricks, general	8 x 4 x 2½ (203 x 102 x 64)

1770	Stock bricks, general	8½ x 4 x 2½ (216 x 102 x 64)
1776	Stock bricks, general	8½ x 4 x 2½ (215 x 102.5 x 65)
early 18th	31 Castle Street, Hereford	9¼ x ? x 2 (23.5 x ? x 5.2)*
mid 18th	30 Castle Street, Hereford	9 x ? x 2 ¾ (23.2 x ? x 7)*
Late 18th	Liverpool, Rodney Street	9½ x 4½ x 3 (240 x 114 x 76)
Early 19th	6 Castle Street, Hereford	9¼ x ? x 3 (23.5 x ? x 7.5)*
Early 20th	Colwall brickworks	9 x 3¼ x 4¼ (260 x 79 x 110)
1965/69	British Standard	8½ x 4 x 2½ (215 x 102.5 x 65)
	*broken bricks	

Bromyard: There are six sites shown, two of which, the Linton/Stream Hall (HSM 34697) and the Sheep Street (HSM 32284) were in Bromyard both in a bigger way of business. Richard Booth, a brick and stone mason had Thomas Brown, brickmaker, as his lodger. Frederick Wall lived at the Toll House on the Downs and was a brickmaker's labourer. Nye and John Pressey living at Downside were brickmaker and labourer respectively. Another labourer was Henry Bannister living in a cottage on the Worcester Road. In Sheep Street, Thomas Watkin and William Green Snr. lived, both brickmakers, while George Green, another, lodged elsewhere in the same street. Edward J. Halford and John H. Paine were brick managers living in High Street, and Joseph William, another brickmaker, lived in Back Street.

Workers at the Bromyard Tile Works believed to be in the 1930s (Bromyard & District History Society Archive)

The Brick Tax 1784–1850

The brick tax was introduced on 1 September 1784 by William Pitt to pay for the American War and there was much opposition to it. It continued until 1850, however, when after a series of Parliamentary commissioned reports it came to an end, no doubt as a response to the need to build more houses and railways. The tax is an excellent source of information about building for the years that it was in operation, the yields from the tax directly mirroring the rate of house building, expansion of the railway network and rise in population.

Bricks were taxed at first at 2s 6d per 1,000 and tiles at 3s per 1,000, but this was quickly raised and by 1805 the rate was 5s 10d and 5s 8d per respectively.[33] Drains were not charged if stamped with the word DRAIN. The tax was assiduously collected, the Excise inspector visiting every brickmaker daily and counting every brick made, but none of the records for Herefordshire appear to have survived.

In 1847, just before the brick tax was repealed, £625,507 was collected nationally, representing 5% of the total Excise duty raised for the year and 1.2% of the total government income.

The brick tax brought a change to brickmaking. The duty was about one third the cost of the brick, but for small brickmakers it was proportionally more. In addition brickmakers now had to register with the Excise, be visited daily by the inspector and undertake a complicated process of being overseen and paying tax. This deterred many of them and led to a certain concentration of the industry into fewer firms.

The bricks on the left-hand side at Brick House, Marden, being larger, may have been made at the time of the Brick Tax

It also meant that works may have aimed to produce bigger bricks, in order to pay less tax. Certainly the mid 18th- to early 19th-century houses in Castle Street, built at the time of the brick tax used bricks that measured 240mm x 70mm. These are much larger than the 1660 Price's almshouse bricks of 240mm x 35mm, or indeed the early, pre-tax 18th-century house No 31 Castle Street which is built of bricks measuring 235mm x 52mm. However bricks were becoming deeper in size *before* the brick tax, presumably because of economies of building and making. We have not found any obvious 'tax avoidance' bricks in Herefordshire.

In 1846, of the 53 tax collection points, Hereford was ranked 46th in terms of tax collected, only Exeter, Bath, Barum (Barnstaple), Plymouth, Mid Wales, West Wales and Cornwall paid less. Hereford collected £2,023, which was a mere 3% of that levied in Surrey.[34] On a per capita basis, however, the tax raised equated to that in Worcestershire and Shropshire. Cheshire's per capita levy was larger, either meaning that the county was developing faster and/or exporting bricks out of the county.

The brick tax showed that bricks were made at a steady rate in the first half of the 19th century, between 5 million and 7.5 million a year in Herefordshire overall, the equivalent of between 250 and 340 small terraced houses a year. The census shows that in 1841 there were 92 houses being built, and 69 in 1851 while population was increasing at about 200 people per year. In 1841 the average number of people per house was 5.1, which would mean about 40 houses per year were needed to accommodate the rising population. These figures show, therefore, that larger houses and replacement houses were also being built during this period, as well as ranges of farm buildings.

The brick tax pre-dated the Victorian terrace and a slightly faster growth in population.

High House Farm, Preston Cross, an example of a black and white farmhouse added to in the second half of the 19th century

Burghill: John Probert, foreman in brickyard and his brother William Hopton lived at Brickyard, right on the premises of HSM 32417.

Burrington: Which of the two brickyards here (HSM 31587 and 32419) these two men worked for it is not possible to say. William Davies was a brickmaker, and Thomas Davies a labourer.

Cradley: There were two men here, William Davies a brick drawer and Joseph Clement Head a brickyard boy. There is one site listed, Brickclose (HSM 32426).

Eardisley: David and Ann (née Burgoyne) Pollard's children were born between 1853 and 1862 as shown on the census. During this time David changed his job from a brickmaker to a brickmaker's labourer, and lastly becoming a master brickmaker. A relative, Thomas Pollard, worked in a brickyard at Kington. Similarly William Morris, a brickmaker himself of Hobby Lyons, and the son of John also a brickmaker, had married Mary Ann, (said to be a minor at the time). They had nine children, and William described himself at their respective births as a brickmaker, a master brickmaker and a journeyman brickmaker (twice). After this he gave up making bricks and

Eardisley: a brick house from the late 19th century

became a farm labourer. Also at Hobby Lyons was George Smith, a foreman pipe and brickmaker, married to Elizabeth and whose first child was born in 1889. In the census William Morris, living at the Common, is shown, but none of his children, who presumably had left home by 1881. There are five entries in the *Gazetteer*, of which Eardisley Brickworks (HSM 31565), is the largest.

Ewyas Harold had the Pontrilas Brickworks, (HSM 11864), sited on the present A465 (now a different business), and Cornelius Hunt, brickmaker, lodged at the Brickyard, while Sam'l Wilmot, a brickmaker lived at Penland Cottages.

The only brickworks at **Grendon Bishop** (HSM 18475) was found from the Tithe Map of 1840, so John Allen, a brickmaker living there at the 1881 census, could have worked here if it was still going 40 years later.

The Old Brickworks (HSM 22485) at **Hampton Charles** may well have supported George Capewell, brick burner, of Luke Lane Cottage, together with David Abbots, Samuel Gunthorpe and Thomas Gunthorpe, his stepsons, and all brickmakers of Holloway Common.

We have no knowledge of a brickworks at **Hatfield**, however Jonathan Hunt was a brickmaker who lived here in 1881.

Hereford: George Davies, a brickmaker's labourer, lived at 86 Widemarsh, Newtown. He could have worked at one of the Barr's Court Brickworks (HSM 32327 or 26973) alongside Joseph Wallis, brickmaker's foreman, Benjamin Baldwin and George Treherne, labourers, all living at the Barr's Court Brickyard. At 9 Barr's Court Road was James Ballinger and his sons Charles and James, all labourers. Thomas Brooks, another labourer, lived at 7 Barr's Court Terrace, and another 'labourer in brickyard', John Smith lived at 10 Brighton Terrace.

It is more difficult to pin these next people to a particular brickworks: Alfred and Willie Skinner, brothers and 'labourers in brickyard' lived at 29 Burcott Row. John Arrowsmith, a brickmaker, was at 39 Catherine Street, and another, Lewington Sweetman lived at 12 Court Friars, while at No 7 was John Sheward, brickmaker. John Williams, another labourer lived at No 1, Court 2. William Bayliss was a hotel keeper and brick manufacturer employing eight men and one boy. William Hill was a brickyard engine driver living at 11 Monkmoor Street. (See also Holmer and Tupsley).

Brick in Hereford City

Hereford is the largest centre of population in the county. In 1664 the Hearth Tax listed 718 houses in the city. Between this date and the time of the first census in 1801 the population doubled to 6,828 people, representing an increase on average of five houses a year. Some of the houses made in this period were built in stone or wood but many were made of brick. Examples can be seen in all the main streets of Hereford inside the city walls, for example in Castle Street, Broad Street, Church Street and the Cathedral Close, with a few in Barton Street just outside the city walls on the west side. Those within the city walls are likely to be re-constructions on the site of earlier houses that had been made of timber, wattle and daub. The bricks that made these houses were hand made clamp fired bricks probably using coal, most likely brought in from the Forest of Dean.

Between 1801 and 1901 the population of Hereford almost trebled to 18,122 people. This represents about 3,624 houses, being on average 36 houses a year, an expansion which can be tracked on maps. As early as the tithe map of 1843 there were houses lining the eight roads that lead into the city. By 1885 spaces between these roads had been filled with regular rows of houses built of bricks from the large brickworks such as Holmer and Hampton Park.

Today the city extends for about 2 km beyond the city wall in all directions and the population is approximately 55,300. The house building rate is about 225 houses per year.

The earliest known surviving bricks in Hereford are at Price's alms-houses in Whitecross (Plates 12 and 13) dating from 1665 and the Sack Warehouse on the south bank of the river, next to the Saracen's Head Inn, dating from the early 18th century, possibly 1732 (Plate 16). The initials FJ carved inside the building may be those of Francis James, bricklayer.

Holmer: The big Albert Brickworks (HSM 32329) and the Victoria Tileworks (HSM 9414) had not long been in business, and these people living in the parish, may well have worked there. William Davies lived at the Rose Garden Inn, Edwin Sweetman at Crossway Road, both labourers in brickworks. Edwin may have been related to Lewington Sweetman, referred to under Hereford. George Allen, brickmaker, and Beaufoy Alcock, labourer, lived at the 'Cottage near the Bridge'. James Addis, labourer in brickworks lived at the Gasworks Cottages and on Grandstand Road lived James F. Preece, brickmaker, Joseph Pritchard, brick labourer, and his sons, Thomas and James, both brickmakers. At the Albert Cottage was the residence of George Whitney the foreman. This is probably one of the houses which still exist, though much altered, on the site of the Albert Brickworks. The timekeeper was James Price living at Gladstone Place.

Hereford: houses at Barr's Court

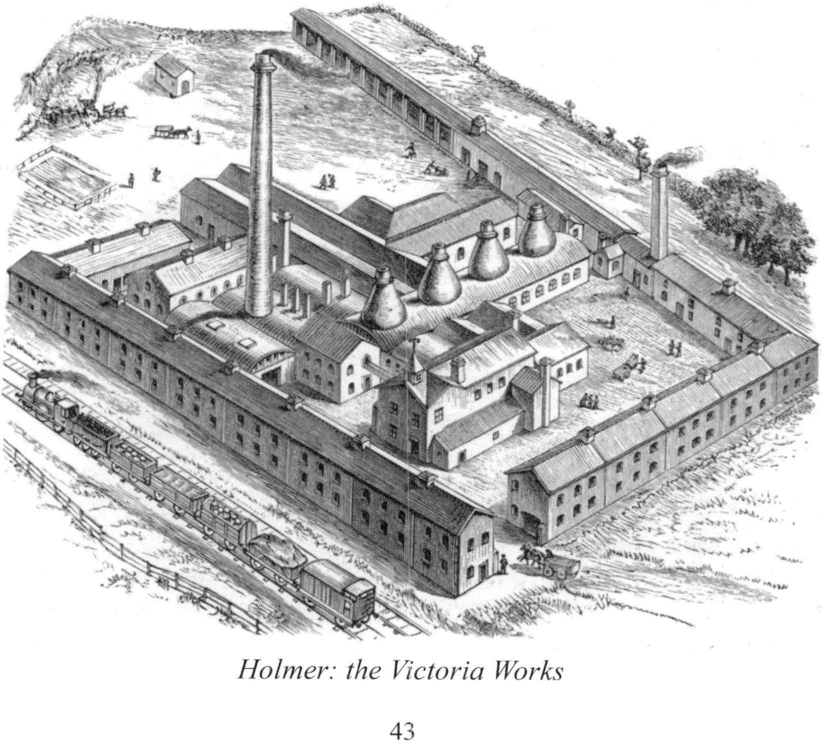

Holmer: the Victoria Works

George W. Wilson of **Kentchurch** probably worked at Pontrilas Brickworks in Ewyas Harold (HSM 11864). He was in charge of 20 labourers.

Kington: There were a number or brickworks in Kington. From the 1881 census we have: Thomas Pollard and his wife Theresa who worked as brickmakers at the Woodhouse Brickyard. One wonders if these Pollards were related to the Pollards of Eardisley. Henry Williams another brickmaker of Mount Pleasant. William Powell, a brick- and tilemaker, lived at 41 Duke Street. He also appeared in Kelly's 1879 *Directory* at 13 Headbrook. At the same address was John William Hurds a brickyard labourer in September 1876. A long established building firm, William Powell & Sons, existed in Victoria Road, Kington, until the 1980s. Also living in Duke Street, No 2, was John Evans, a brickmaker. Thomas Bowen, farmer and brickmaker, was proprietor of the Kingswood Brickworks (HSM 32335). They manufactured bricks, tiles and drainpipes, and also appeared in Kelly's *Directories* for 1879 and 1885. Thomas Pollard, again, was a brickmaker when his first child was born in May 1881, and he lived at Kingswood. Another Kingswood resident, at the Brick Kilns this time, but probably the Kingswood Brickworks, was David Beddows a journeyman brickmaker. George Seaborn was a labourer, the son of John, a millwright, when he married Ann, née Weaver, on 21 October 1844. They had six children and when registering the births of the first two in 1858 and 1860 he was a brickmaker at Flintsham Brickworks (HSM 32487). However, when he registered the last four, who were born between 1863 and 1872, he had moved to the Brick Yard at Titley (HSM 32461) where he was variously described as a master brickmaker, and a journeyman brickmaker. Thomas Hoskison, married to Eleanor, née Long, was a master brickmaker. He lived at Sunset, where his first child was born, but had moved to Elsdon at Lyonshall when the second came along.

Kington: 2 Duke Street, the home of John Evans, brickmaker

Kington: 13 Headbrook, the home of William Powell, brickmaker

Kington: 41 Duke Street, an earlier home of William Powell, brickmaker

Ledbury: There are seven known brickyards in and around Ledbury, but a paucity of workers in the census. William and Walter Dobbins and William Halwood are all described as brickmakers living at The Cottage, Bye Street. In New Street is Joseph Hodges, also a brickmaker, and George Elsmore lived in Newtown. At the time of the census his father, also a George and also a brickmaker was visiting.

Leintwardine: For once, it is possible to state categorically where all six resident brickmakers worked, for they all lived at the Brakes Brickyard (probably HSM 32339). John Amphlett, was the likely owner, described as brick- and tilemaker with twelve men. Benjamin Micklewright appears to have been unemployed since he describes himself as 'formerly labourer in brickyard'. However the lodger Alfred Micklewright was an employed labourer there, and another lodger, John Kibble was a bricklayer. At the Forge, adjacent to the Brakes, were two labourers at the brickyard, John Davies and John Hammonds.

Leominster: We have recorded seven brickyards in Leominster, but there are only fourteen people claiming to work at a brickyard. Abraham Tunks was a brick manufacturer and inn keeper. He was the proprietor of the Barons Cross Brickyard (HSM 32340), and his son Leonard was the foreman. Another foreman elsewhere was William Savigar living at Lower Brickfield

Leominster: the home of William Parker at 33 Bargates

Leominster: the home of Richard Pendry at 2 New Street

(unlocated) with his unstated relation John J. who was a brickmaker's labourer. Another Savigar, William, a brickmaker lived at 3 Vicarage Street,. There were more brickmakers: John Rogers at 1 Grange Cottage; Joseph Fletcher, 7 Richards Court; Edward Fletcher, 10 Pritchards Court; Richard Pendry, 2 New Street; Benjamin Bateman, a brick- and pipemaker, 54 Etnam Street; Henry Davies, 11 Vicarage Street and Charles Hunt, 40 Dishley Street. For all these brickmakers and foremen there were only five labourers: George Palmer, 19 New Street; William Parker, 33 Bargates; James Johnson, 41 Dishley Street, another James Johnson of the Brickyard (address not stated) and Andrew Deakin of Barons Cross.

Brick Types

'Stock' or **common** bricks are the common, ordinary brick, the word deriving from the stock board on which the hand made bricks were formed. **Place** bricks are often poorer quality, badly fired or misshapen in some way and these are placed behind the facing bricks. **Facing** bricks are those on show, they may be polished, or organised into a pattern. **Engineering** bricks are higher fired, usually from the Staffordshire Carboniferous coal measure clay that can be burnt to 1500°C without vitrifying and as a result have a higher load bearing capacity. **Flettons**, or London bricks, are stock bricks made of the Oxford Clay discussed on pp.27-28.

Living in **Laysters** were two brickmakers, William Nash of Pole Lane and Joseph Band living at the Brickyard. It is probable that Charles Cooke, although living at Pole in Middleton-on-the-Hill was a labourer at the Laysters brickworks (HSM 31583).

Edward Cook was a brickmaker of **Lower Bullingham**, and another of the same name was a bricklayer, living at Ridge Hill.

William Godwin of **Lugwardine**, brick and encaustic tile maker, lived at Porch Farm and had two brick- and tileworks in the parish (HSM 11862 and 11863). (See the history of the Godwin firm below). Joseph Spencer and Joseph Arthur Spencer lived at Lugwardine (no more specific address stated) as brick burner and assistant brick burner. At Emhill (today Hemhill) lived Thomas Tew, brickmaker.

Laysters: ruined house on the Laysters' brickworks site, probably the home of Joseph Band

The Godwin Family

One of Hereford's leading brick- and especially tilemaking families was that of the Godwins, who specialised in decorative tiles. Their business started in Herefordshire with the arrival of William Godwin in 1849 when he was 36 years old. He had been born in Leicestershire, but had moved to Gloucestershire where he worked in the brickfields beside the River Severn, first as an apprentice and later as a manager. He then moved to Ledbury, a centre of brickmaking during the 19th century. Ambitious, he spotted an opportunity when the tenancy of the brickworks owned by C.A. Mason in Lugwardine came on the market, buying it and moving in with his wife and five children. Two years later the 1851 census shows that the household had increased by a servant girl and a 14-year-old brickmaker's labourer. The family prospered. A relative, Ebenizer Godwin came to the village as a brickmaker's labourer with his wife and baby son, then a few years later Eleazer, William's half-brother, came to help in the firm.

The original factory was in the lane to the north of the church in Lugwardine. All that now remains are cottages built from old bricks, and existing buildings re-developed (Hemhill HSM 11862).

The diversification of the business into tiles was due to the arrival of another Godwin. Henry was William's much younger brother who had leaned this trade at Maw's of Worcester, and he came to Lugwardine when he was aged 24 in 1853. He was said to be a man who could write little more than his name, but with great skills as an experimental chemist and able to carry calculations and formulae in his head. With William dealing with the business details and Henry with the research and development, they seemed to be an ideal partnership.

In the early days the tiles were all six inches square and made from red plastic clay with a white clay inlay. They were of such quality that George Gilbert Scott used them in the 1857 restoration of Hereford Cathedral, and orders came for work being carried out by Pugin, Seddon, Burges, Street and others working on Victorian church restorations.

In 1857 William bought a site at Tupsley which he used for brickmaking, leaving the Lugwardine factory exclusively for tiles. Only four years later in 1861 more space was needed, and William bought another site on a good clay bed at Whitestone in Withington (SMR 11863). It was beside the railway station on the newly opened Hereford to Worcester railway, and only a mile and a half away from the Hereford to Gloucester canal, which stayed in use until 1881. Godwin's Encaustic Tiles (decorated with firing) were made at Withington, whilst bricks were made at Tupsley and 'tiles and sanitary pipes made by patent machinery' at Lugwardine.

The tiles were widely advertised at exhibitions in Britain and abroad. These were the years of prosperity, and William was busy buying the freehold to as much of his property as he could. By this time he was living at Porch Farm, Lugwardine, a few yards south of his Hemhill works in Lumber Lane. Henry was living in a house at the factory site at Withington, and William Henry moved into an exuberantly decorated house on the main road at Bartestree, now the New Inn, where there is still much tile decoration.

The success of the business began to take its toll on William's health, and in 1880 he had a succession of strokes. He took William Henry into partnership and the firm was called W. Godwin and Son. He died on 21 August 1883, leaving the business to William Henry, and was buried in Lugwardine churchyard in an elaborately tiled vault, a copy of the one made by Godwins for George Gilbert Scott. William Henry inherited a successful business which employed about 100 men and a number of women and boys producing tile hearths and fenders with hand-painted pictures of landscapes or flowers.

Advertisements, at this time always carried the rider: 'Messrs Godwin & Son wish it be distinctly understood that they have no connection whatever with any other firm using the name of Godwin and that their only address is as above, i.e. Withington.' This is evidence of a family feud, for Henry Godwin had left the original business in 1876 and set up a rival factory. Every tile carried the Godwin trade mark, and surveys of church floors have shown that W. Godwin tiles were used on more floors than the total by other manufacturers.

William Henry was not a successful businessman like his father, and the firm gradually declined. In 1906 it was sold to G.H. Lloyd and Thomas Pulling who concentrated on glazed fireplaces. At about this time the Lugwardine factory was closed. Six years later the Withington factory was sold to a local farmer and fruit grower, T.E. Davies, and for the first time there was no Godwin on the board. Davies had bought the buildings, plant and goodwill for £8,500, paying £3,000 in cash and the balance as shares in the new company. It wasn't a great success since part of the factory was used for making jams, pickles and bottled fruit.

After a further change of ownership in the early 1920s, an attempt to re-equip the factory was made, but the money ran out, and the firm was liquidated. It was sold to H. & G. Thynne Ltd for £12,500 who began making white tiles there. In 1958 it was bought by H.R. David who changed the name to Hereford Tiles, continuing on the Withington site until 1980 in the original buildings.[35]

Lyonshall: John Herds (*sic*) was head brickmaker at the Elsdon Brickworks, which we have not been able to identify. The Herds family was still in the Lyonshall area in 2002. Peter Klein's research found that William Walters, journeyman brickmaker, was living at Lower Rodds, close to Elsdon, in December 1861. Thomas Hoskison appears at Eldson as a 'brickmaker master' in March 1865; and Joseph Dally or Daly was a pipemaker at the Elsdon Brickyard in 1870, the following year he described himself as a journeyman brickmaker, but by 1875 he seems to have come down in the world and was a labourer in a brickyard. Thomas Jones of the 'Old Bridge Inn, Headbrook', Kington (now 23 Bridge Street), was proprietor in 1876. Kelly's *Directory* reports that Jones was still at the inn in 1879 but there is no mention of bricks, although the same *Directory* entry for Lyonshall still says 'bricks and tiles are made there.' John William Hurds (*sic*) lived and worked there as a labourer at least between April 1879 and August 1883. There is no reference to brickmaking in Kelly's *Directory* for 1891.

Thomas Jones of Westmore, **Mansell Lacy**, was a farmer and brickmaker. This is possibly a case of a man having made bricks for a building of his own, now wanting to make a business of it.

Charles Cooke of Pole, **Middleton-on-the-Hill** was a brick labourer probably at the Laysters Brickyard.

Samuel Savegar of **Much Cowarne** was an 'ag, labr. (brickyard)' and lived at Little Bromtrees. With this rare surname, was he perhaps related to the Savigars (*sic*) of Leominster, q.v. The Burley Gate Brick and Tile Works (HSM 31569) is in the parish and about three miles from his home. However Little Bromtrees today is in Bishops Frome, and he would live nearer to Bromtrees Hall Brickyard (HSM 32481).

At **Norton** by **Brockhampton** lived George Masters, a brickyard labourer, at Sallow Cottage. William Snelgrove was a brickmaker living at the Cottage.

Richard Shepard, a brickmaker lived in the 'Village' at **Orleton**.

Post Office and Shop in Pembridge, the house of Thomas Powles, brickmaker. The west side of this timber-framed house has recently been dated to 1583, the east side is earlier

Pembridge: There are eleven known brickyards in Pembridge. John Morgan lived at Upper Whettons (*sic*) Cottage, and no doubt worked at the Upper Wetton Brickworks (HSM 31577). Eliza Howes was a brickmaker living at Broxwood and must have worked at Lower Broxwood Brickworks (HSM 31578), and her labourer was James Ford of Broxwood Brook Cottage. Thomas Powles was a 'builder and brickmaker employing five men', who lived at the post office and grocers shop in the village, and Richard Leader, a brickmaker, lived at Trap Style.

John Perry was a brickmaker in 1853. William Swain Jnr. is recorded as being a brickmaker in 1841 and 1842 when his children were born, and he appears in Lascelles' *Directory* of 1851 as a brick- and tilemaker. Thomas Tidman married Mary née Swain, probably the daughter of William Snr, and was a brickmaker in 1840. James James, who married Jane Watkins was a journeyman bricklayer who worked at three places in the area: in 1860 at Pembridge, in 1864 at Broxwood, and from 1867 at Bonds Green. James Hackman a journeyman brickmaker lived at New House Broxwood in 1872, and George Clee another journeyman brickmaker, also lived at Broxwood in 1871.

At Trumpet Post Office, **Pixley** lived Thomas V.D. Jones who is unusual for the time in being working class and having three first names. He was a brickmaker's labourer.

St. Devereux: There are three sites in this parish within 2 km of each other. Two are called Brick Close and were probably farmers' brickyards for a specific building. One was at Stock Farm in the east (HSM 32466), the other at Vallets in the west (HSM 32465). In the centre was the Crizeley Brickworks (HSM 31559), a small commercial operation. We have records from 1881 of two brickmakers: John Spencer, a brick burner, who was a lodger at Spring Cottage, and George Morgan of Grove Cottage

Sarnesfield: Daniel Dance lived at the Batch (HSM 31567) in 1881 but by 1891 the Batch Brickworks was described on the Ordnance Survey Map as being disused.

Stanford Bishop: There was a brickworks on Woofferwood Common (HSM 31575) and the census gives John Patterson as a brickmaker living at Three Springs and employing one man. The Kelly's *Directory* of 1891 lists a brickworks owned by Thomas Patterson and Son on Woodford (*sic*) Common.

St. Devereux: Brickyard Cottage

Charles Green of **Staunton-on-Wye**, brickmaker lived at The Old School House.

Stoke Edith: From the records above, under Early Herefordshire Brick-workers, brickmaking had been going on here since at least 1696. In 1881 there is a record of James Hinton, brickmaker, who lived at Newtown.

William Weaver of **Tarrington**, brickmaker, lived at Saillon. He may have been employed at the Brick Kiln recorded on the tithe map (HSM 18222).

Tedstone Wafer: William Shaw and his son George, lived at Pine Tree Cottage. There was a brickworks here called Prince Albert Works, shown in Kelly's *Directory* of 1891 (HSM 31582).

Tupsley: It is most probable that all the workers listed as living at Tupsley were employed at the Hampton Park Brick and Tile Works (HSM 31564). Most of them lived in Old Eign Hill, within a quarter of a mile of the main works. George Preece was brickworks' manager, and his relation, Alfred was a brickmaker, both of whom lived in Old Eign Hill at the Brickworks. Living at the Whalebone Inn was John Downes, brickmaker, and at 2 Franchise Stone Thos. Norman was another. In Walcutt Cottage on Old Eign Hill was George Gregg, brickmakers' labourer, and just visiting him was a brickmaker, William Shermer. At No 2, the Terrace, was Thomas Griffiths another brickmaker's labourer, and with the address of only Old Eign Hill were three brickyard labourers: Walter Smith, John Field and William Williams. Another labourer in a brickyard was Francis Preece who lived at 1 Wellington Place, Old Eign Hill.[36]

Bonds and Patterns

There are many different ways and patterns of laying bricks but the three most common are English, Flemish and Stretcher bond. English bond consists of alternate layers of headers and stretchers, with a quarter brick to balance at corners (see Plates 17 and 19). Flemish bond consists of alternate header and stretcher in each row, with a similar quarter brick at the corners (see Plates 20 and 21). Stretcher bond has all long sides facing and is used for cavity walls, but was also used in the past for some garden walls (see Plate 24). The earliest bond is supposedly English, while Flemish bond began to predominate from the mid 18th century and stretcher bond from the mid to late 20th century.

There are some more unusual bonds. Header bond is for walls of one brick thickness, and in Georgian times was popular for house fronts, especially in grey bricks which had been relieved by red brick quoin, reveals and window arches. Flemish Garden Wall bond is a cheaper version of Flemish bond, with two stretchers and one header along the row. English Garden Wall bond is yet more economical of bricks, being three rows of stretcher and one row of headers. From these bases almost any design of headers and stretchers can be made, and they carry delightful names such as rat-trap bond, flying bond, irregular bond and Quetta bond, to quote only four.

Anthea Brian's survey of bonds across England found that Flemish bond was used in the front of houses and in main streets—it was seen as the 'posh' bond. Our observations in Herefordshire have found that most Victorian houses are built in Flemish bond, and most Georgian and earlier houses and walls in English. However apart from late Victorian town houses, it is rare to see a perfect example of either of these types, and most buildings, particularly early ones, seem to use a mixture of bonds.

Internal walls of housing were built 'one' brick wide, which with English bond meant two bricks laid side by side for the stretcher row and one brick on top for the header row, making the wall one brick length wise wide, i.e. about nine inches. **External** and **joining** walls were built either one and a half bricks wide, which meant a stretcher course backed by two headers lying parallel making it a 13.5 inch wall or a two brick wall. This was either two headers laid short end to short end, or headers with a stretcher facing both front and back, making an 18-inch wall.

In English bond the aim is to make no continuous vertical joints, so the wall is strong, but uses slightly more bricks than Flemish. Flemish is meant to make a neater wall, as bricks inevitably are different sizes and the appearance can be more easily controlled with the long, short sequence of Flemish, but the wall has the potential to be weaker because continuous vertical joints can appear at the end of walls. English bond is slower than Flemish bond to build.

Cavity (half brick) walls are the fastest to make, a practiced bricklayer today can lay over 10 square metres a day with this type of bond. The advantage of cavity walling is that they provide damp proofing and insulation and use less brick, but they are not so attractive. Because of this when cavity walls were first made in the 19th century people used Flemish and English bond, 'snapping' the headers to achieve the cavity.

Further Information

Very many members of the public responded to an article in the local press and interviews on local radio in 2002. We have been able to trace some of the brickworks they referred to and their comments have been included appropriately. However some we were unable to be definite about, or reference was to buildings and these are included here.

Norman Davies of Malvern reports: Hill House, **Ledbury** was made with Davies bricks stamped JCD, also Pixley Rectory and houses on Bye Street. When his grandfather died in 1931, and Norman was 15, clay came to Ledbury on tractor engines brought down from the railway. Some clay was imported, and the steam engines frightened the horses. (See also next entry.)

Steve Glennie-Smith of Ledbury said George Hill, **Ledbury**, continued for some time as a business after the brickworks closed. They were an ironmongers in the premises now occupied by Ceci Paolo at the Top Cross, and an agricultural merchant, later taken over by Dalgety Arable. Dalgety occupies the same premises on the Bromyard Road, and it is possible this was the site of the brickworks. The bricks are very soft. There was also a Station Brickworks, Ledbury, which implies it was sited near the railway station, but he doesn't know where. By contrast these bricks are very hard. J.C. Davies, **Ledbury,** was on the site occupied by Ledbury Preserves, Little Marcle Road.

(This site was originally Ballard's Brickworks and is HSM 32792, with its claypits on HSM 35553.)

Margaret Wilce of Walford told us: 'Mother used to tell me about the cottage industry before the 1930s of a husband and wife team who made brick on **Boar's Hill**. They dug out the clay, there were huge ponds where the clay had been taken, it was not until the 1960s and 1970s that they stopped working and the ponds filled in. There were also a lot of limekilns and quarrying carried on in a small scale manner'. She asked her 92-year-old uncle, James Webb, and he confirmed that bricks were made at Brickyard Cottage.

Major Robert Barnes of Lyde owns property in **Gwynne Street, Hereford**, where Queen Anne bricks have been used, and where it is possible to see other bricks used in later re-builds, and there are parts built in 1905 and 1906 using Hampton Park Bricks. There was a brick kiln at **Shucknall** at NGR SO 594 429 which closed in the 1800s. He filmed the **Holmer** chimney stack being blown up in 1968/69.

Mrs. Webster told us that Hopton House, **Canon Frome**, where she lived at one time, was built on a brickworks. It was full of bricks, there were kilns in the front garden, and pits in the back garden in 1935. The bricks were made for the estate.

Frank Barton of Rotherwas said there was a brickworks in the wood on the right hand side of the road as you leave **Hereford** on the Straight Mile (Holme Lacy Road). Clay was taken down to the river on sledges. There was a brick wharf on the bend.

Plate 1 A Hampton Park Brick: Devonian mudstone

Plate 2 Bricks at Hampton Court, London: London clay

Plate 3 Shucknall Brick: Silurian mudstone

The Mould

Preparing the clay

Filling the mould

Squaring off

Plates 4-11 Brickmaking demonstration at Ludlow Food Fair, May 2003

Continuing squaring off

The brick is knocked out

The brick

Bricks ready for firing

Plates 12 and 13
Price's almshouses (above) and close up of the clamp fired brickwork (left) dating to 1630-1660

Plate 14 (below)
Possible 15th-century brickwork at Hellens, Much Marcle

Plate 15 Early brickwork of c.1732 at the Sack warehouse, along River Terrace, south of the Wye in Hereford

Plate 16 Now the New Inn at Bartestree, this was once the home of Henry Godwin, founder and proprietor of Godwin's Tiles. Note the decorative Victorian brickwork

Plates 18 and 19 (above) The loose brick found in Hereford Town Hall after a fire and part of the fine English Bond walling of the Town Hall

Plate 17 (above) English Bond in multi-coloured brickwork in Chandos House, Hereford
Plates 20 and 21 (lower right) Victorian era Flemish Bond on the Meadows Memorial Hall together with decorative brickwork and an earlier version at 6 Castle Street, Hereford

Plate 22 A form of Garden Wall Bond: Five rows of stretchers and one of headers, this in East Street, Hereford

Plate 23 (right) A form of Flemish Bond with 2 stretchers, then a header — only every now and then there's just one stretcher, or three, or even more! This intriguing display is on a building near the back of Hereford Town Hall

Plate 24 Most modern brickwork is built in stretcher bond (right)

Plate 28 Decorative work on 83 Green Street, Hereford

The joy of brickwork
Plate 25 (above) One way to form a chimney stack, this on the Social Services building on Barr's Court Road, Hereford
Plate 27 (below) Recent Stretcher Bond brickwork in poor quality Common bricks abutting earlier Flemish Bond brickwork. This is at the rear of Woolworths in East Street, Hereford

Plate 29 Modern brickwork in Deen's Court, Hereford

5 Brick Buildings in Herefordshire

Nine sample brick buildings are described below as representatives for each century from the 14th century onwards. In three of the centuries (17th, 18th and 20th), we have given two interesting buildings. Since brick building in the county was rare until the 18th, our choice has been somewhat restricted until the 19th century.

14th/15th Century: Hellens, Much Marcle

Hellens estate was the centre of the manor of Marcle Audleys or Marcle Magna. The brick dovecote, (HSM 482) is 15th century and the oldest part. It used to be the tower of the original castle, but in the first half of the 17th century was renovated and turned into the dovecote. Built using English bond, the bricks are orange, variably fired, and narrow at about 240mm x 112mm x 45mm. Straw markings can be seen, showing that the green brick was laid

East elevation of Hellens

on and covered by straw when being fired. Similar bricks can be seen on the front of the house and elsewhere. Other parts of the building are later and use deeper brick in Flemish bond.

From the 14th century Hellens belonged to the Wallwyn family. In 1451 the manor house was renovated and built in brick. Workmen from Flanders were brought over to teach local labourers the brickmaking skill and the bricks were made from the clay from a quarry that eventually formed the pond behind the house, though this has been enlarged through time (see Plate 14). The brick built quadrangle, and the crenellated tower defending the main entrance, all formed part of this renovation. In building in brick, Thomas Walwyn, the lord at the time, was being entirely fashionable. The age of castles was over and the age of the country mansion just begun.

The house has undergone renovations throughout the following years. A period that clearly leaves its mark is that of Fulke and Margaret Walwyn who inherited Hellens during the reign of James I (1603–1625). The dovecote was renovated and their initials FMW inserted (see photograph below). The medieval look was removed and a Jacobean aspect replaced it.

The Dovecote at Hellens

In the late 17th century a Walwyn heiress married the London cloth trader John Noble whose notebooks and accounts survive. In the 18th century the south wing was burnt down and rebuilt, using bricks again made on site. For this last renovation, there exists an agreement between Edward Walwyn and James Williams dated 18 March 1791, whereby J. Williams agreed to produce between 40 and 50,000 'well burned sound building bricks, each brick when burned to be of the following dimensions—in length nine inches and a quarter, in breadth four inches and a half and in depth or thickness two inches and five eighths of an inch [237mm x 115mm x 65mm] ... For each thousand ... shall be paid the sum of 8/-'. He agreed also to make between 9 and 10,000 bricks in the same clamp 'in length 11 inches, in breadth seven inches and five eighths of an inch or thereabouts [289mm x 195mm] and of a proportionable substance and thickness'. These must have been roof tiles. He was to be paid 12s per 1,000. This must have been a very large clamp, something similar to the ones illustrated in Payne's *Microcosm* (see pp.4-5).[37]

16th Century: Brick House, Marden

Though several 16th-century houses with brick survive in Herefordshire, they are all timber-framed with brick infills that are later additions. Brick chimneys were built at this time on bigger houses and it may be that some originals survive. The only 16th-century buildings using brick in the structure that we have been able to trace are Poplands in Mill Street, Leominster, which supposedly has a brick frame, Tudor Cottage at Lingen which has a brick plinth and the Shambles Barn in Kingsland which has a brick frame.

Brick House, Marden, originated in the 16th century but has many later additions. It is a timber-framed building, the main block and cross-wing both have exposed timber-framing, of which the main front has been refaced in brick. On the 1720 Coningsby map it is shown as a north/south block with two small buildings in a courtyard. On the other side of the road is the Manor House, the same as today, to which Brick House probably belonged. On the

The Brick House, Marden

1840 tithe, Brick House is shown as a smaller block than it was in 1720, perhaps the reason why it looks cut short today. It was owned by Thomas Davies who owned only the six small fields which surrounded it, all of which were orchards and hopyards, though he may well have rented more land.

The bricks for the house were dug from the field to the west, where there are distinct banks. The change in brick size on the west elevation is an interesting feature. It is supposed that the larger ones were used during the time of the Brick Tax.[38]

17th Century: Price's almshouses, Whitecross Street, Hereford

Price's 'hospital' (see Plates 12 and 13) was not a hospital in the modern sense, it was established to provide accommodation for poor people of 'good' character, the criteria for 'good' being determined by the trustees. The charity was founded in 1604 by William Price who made his money as a merchant tailor in London.

The chapel and original 12 houses were built in 1636 which were immediately inhabited. Towards the middle of the 19th century, the houses were remodelled, enlarged and re-slated, and a garden was attached to each house. In 1914 the charity's income stood at £531 which provided 12 'poor men' with 10s a week and coal, and if married their wives could live with them. The Vicar of the Holy Trinity received £20 a year for conducting services twice a week in the chapel. The almshouses are still lived in today and more have been built in the 20th century to the rear, making a complex of small houses around a courtyard.

The lower levels of the almshouses are built of large blocks of grey Devonian sandstone. Chisel or axe marks can clearly be seen on the surface of this good quality freestone and it possibly derived from Blackfriars monastery or the then largely defunct Hereford castle. The bricks that form the main part of the building are very narrow, 35mm x 240mm, and vary in colour, a typical feature of clamp fired bricks. The chimneys were replaced in 2004, sadly none of the bricks were kept as they were very soft and 'just crumbled away'.[39]

17th Century: Castle House, Eardisley

Castle house is a handsome, early brick house dating to the latter part of the 17th century. It lies immediately east of the castle mound, is square in plan and set on a stone plinth. The bricks are hand made, long and narrow, and laid in English bond with some blue brick dressing. Today the house is a 'double pile' house, because a twin building was added to the rear of the main block in the 19th century. A further 17th-century, two-storey, brick building with a large chimney lies to the north. This was probably built as a service block to the main house and housed the dairy, bake house and kitchen. Castle House

Castle House, Eardisley

must have been built by a fashionable client with plenty of money, no doubt the lord of the manor. The brickmakers worked on site, perhaps excavating their clay from the ponds that originated as the castle moat.[40]

Early 18th Century: St. Mary's Church, Tyberton

Built in 1721 the church contains a number of memorials dating from the time just after the Civil War (1668). It is a rarity as a brick-built country church.

The Lord of the Manor, William Brydges was a London lawyer who left many of the matters at Tyberton in the hands of his father, Francis. There does not appear to have been an architect for the church, but principles laid down by William were followed. The mason in charge was Robert Pritchard of Clehonger, and the church was built in brick, a material in vogue at the time, the bricks being made on the estate. Three sizes of bricks were made: common brick, common bricks 'made larger in every way' for rubbing (i.e. polishing for special effect), and long bricks for the arches of the window. The prices paid were 4s 6d, 5s and 7s per thousand respectively. The first forms the material of the walls, the second was used in the porch and the third no longer exists since the Victorians wanted to 'Gothicise' the windows and replaced the arches.

By the end of 1719 the body of the church and tower had been built, but there was a dispute with the Dean and Chapter of Hereford Cathedral over the chancel which delayed the completion. The Duke of Chandos, a cousin of

St. Mary's church, Tyberton

William Brydges, recommended John Wood, the well-known Bath architect of the Crescent and the Circus there, to design the Chancel. The brickmaker was a Mr. Lawrence who is recorded as burning bricks for the same part of the church.[41]

18th Century: Moccas Court, 1781

Moccas Court is a red brick mansion with a basement and a hipped slate roof (see photograph on p.34). For reasons that are not clear, plans were commissioned from the Adam brothers, but built to a design by the Gloucestershire architect, Anthony Keck in 1755–1781. The bricks were manufactured on site in Brick Kiln, now a wood, situated behind the church, and Thomas Williams is recorded as the brickmaker in 1776. The bond is Flemish and the Bath stone used in the sills and architectural features was probably brought up the river, which is immediately below (to the north) of the house. (This property can be visited on certain days in the summer.)[42]

19th Century: Railway Viaduct at Ledbury, 1861

The Ledbury viaduct is Herefordshire's greatest monument to the Railway Age. The viaduct spans the Leadon Valley, and has 31 arches, is 19m high and 300m long. The arches are round-headed on slender piers, with continuous hood mounds of dark brick. Twentieth-century railings and refuges have been added to the original building of projecting cornice and plain parapet. The viaduct was built for the Worcester and Hereford Railway Co. by the firm of Ballard and Barrey.

The design is based on the railway bridge crossing the Severn in Worcester. Railway engineers favoured the arch since it was well understood and it was used in almost all brick and stone bridges and viaducts. The first brick railway bridge was opened in 1826 on the Stratford-upon-Avon and Moreton Tramway and the first brick railway viaduct in 1836 on the London and Greenwich Railway. The Greenwich viaduct ran for 3.75 miles, and was developed as a way of avoiding a large number of level-crossings in the London suburbs. It was made of grey brick and had 828 arches which were 22ft high with an 18ft span. Other brick viaducts followed.

The last brick of the last arch of the Ledbury viaduct was laid on June 12 1861. A single track from the west was complete enabling guests to arrive by special train for the opening ceremony. This single track was later doubled, and then reduced again to single track in 1982. The viaduct was the greatest engineering feat in the county, after that of the railway itself. It is built of an estimated 5,000,000 bricks that were dug and baked on the site by Robert Ballard's firm, the well established brick manufacturing company in Ledbury (see HSM 32792). Robert was also a partner in the engineering firm that built the bridge. The viaduct is on private property and cannot be visited without the permission of the owner.[43]

Ledbury Viaduct

20th Century: Hereford Town Hall, 1902

After the demolition of the Old Town Hall in High Town, the council's work was conducted in a succession of premises in the city, but as its responsibilities grew it needed a new home with modern facilities and accommodation. In May 1900 the Misses Johnson offered to donate a row of three buildings in St. Owens Street valued at £5,000 specifically for the Town Hall. This was gladly accepted and the adjoining premises of the Young Women's Christian Association (YWCA) was bought in addition.

The building was designed through a competitive tender open to all architects in Britain which was won by Mr. H.A. Cheers of Twickenham. The building costs were estimated at £18,500, which with heating and lighting at £1,500, the architect's fees of £2,000, and furniture of £3,000, brought the total costs to £25,000. As ever unforeseen expenses pushed the price of the building and furnishing to £28,135, which together with the purchase of the YWCA premises at £3,276 brought the total to £31,411, causing something of a scandal.

The foundation stone was laid by HRH Princess Henry of Battenburg on 13 May 1903 during a visit to decorate local heroes of the Boer War. The Household Cavalry, who had come up on the train from London, provided

Hereford Town Hall

her with a colourful, and for Hereford, rather exotic escort for the occasion. Thirteen months later, in June 1904, the Town Hall was opened by the Mayor, Alderman Beddoe, with yet more pageantry and splendour. There was a religious service, led by the Bishop and the Cathedral Choir, the Mayor opened the door with a golden key, the National Anthem was played, the Cathedral bells rung and the city flag hoisted. Then followed a sumptuous lunch of 'first class character'.

A local builder, W.J. Bowers of Bath Street, Hereford had won the contract to build. Over the years, however, the construction details have been mislaid, and it was only during the writing of this book when a small fire was started by a dropped cigarette that brought the Fire Brigade, that we were able to establish with certainty the origin of the bricks. In the aftermath of the fire, a small hole with loose bricks was revealed in the walls of a basement room. Two samples were taken, and both were stamped with The Hampton Park Brickworks. Their colour and sizes were compared with bricks in the structure and were found to be sufficiently similar to confirm their use throughout. This, added to the fact that W.J. Bowers was also the owner of Hampton Park Brickworks, we felt was conclusive (see Plates 18 and 19).

The Town Hall is a Grade II* listed building with a slate roof and faced with terracotta mouldings with a striking baroque frontage. It is of symmetrical design with false balconies and the city crest in the centre. The Hampton Court bricks are remarkably uniform in size and colour and have been laid with minimal bands of mortar, so the effect is almost of terracotta throughout.[44]

20th Century: Secondary School for Boys, Widemarsh Street, on the corner of Blackfriars Street, 1912

The Education Act of 1902 allowed county councils to provide secondary schools for both sexes, but separately. Hereford was quick off the mark, and by August 1903, a committee set up for the purpose had determined that a secondary school for boys should be established. However it took the committee till June 1906 to report back its recommendations. These were to provide a school for 150 boys from the county 'centrally situated, ... and provided with laboratories and workshops and other accessories ... for imparting a thoroughly sound, modern and practical education for pupils.' There was to be no provision for boarders since it was felt that pupils from the county could stay at a lodging house in the city. Boys were to be admitted at the age of eight years, and only in exceptional circumstances could they stay beyond 18.

It was another five years before the plans were submitted by the County Surveyor, Mr. G.H. Jack, and the tender of Wilks and Son was accepted at £7735 to build it. The furnishings and land brought the cost up to £10,000,

The Secondary School for Boys in Widemarsh Street

and it was felt that this was a reasonable price to pay to give 150 boys a first class education.

The accommodation consisted of a heating chamber and caretaker's stores in the basement; Assembly Hall, six classrooms, cloakrooms, Masters' Rooms, lavatories and cycle stores on the ground floor; and lecture room and rooms for teaching science and art, and a dining room 'for boys who have come some distance' on the first floor The buildings incorporated the latest improvements in school design being well ventilated and lit. The school is a wonderful showcase for Hampton Park brick which forms the main fabric of the building. It is finished off with Bath stone dressings and a slated roof, and inside the floors are of pitch pine blocks and granolithic paving.

The school was opened in 1912, the fees were fixed at £6 a year and 5s towards the Sports Fund. At the time plans were in progress to provide a similar school for girls.[45]

20th Century: Municipal Housing in Hereford

Urban conditions of the mid 19th century for the working classes were cramped, damp and insanitary. These conditions along with an inadequate diet resulted in high mortality rates. The Public Health Act of 1848 established local boards of health, and this was followed by Housing Acts in 1851, 1868 and 1875. Chadwick's Report on the Sanitary Conditions of the Labouring Poor of 1884, showed that these acts had had little effect. More enlightened employers however, for reasons of philanthropy and to improve productivity, established better conditions for their workers.

The Garden City Association had been formed in 1899 (later to become the Town and Country Planning Association). This led to the establishment of Letchworth Garden City from 1902 and Welwyn Garden City after the end of the First World War. In the meantime the Housing, Town Planning etc Act of 1909 gave permissive powers for Local Authorities and other landowners to prepare schemes for urban growth. This was followed after the War by an act of the same name in 1919, which led to 'a series of experiments in State intervention to increase the supply of working class houses'. A new standard of working class house was established of three bedrooms, a kitchen and bathroom and built to a density of not more than 12 per acre.

The Housing and Working Classes Act of 1903 was used by the Hereford City Council to buy land at Barrs Court on which to build houses. The land was leased to Hereford Co-operative Housing Ltd., a registered industrial society. A government grant was obtained and the Hereford architects, Groome and Bettington, asked to design the houses. Eighty-five 'cottages' were built, which gave accommodation for over 400 people. The design was advanced, being based on the principles pioneered by Ebenezer Howard and the Garden City Movement established only five years before. As a consequence the development received a Garden City name, though in fact it was not the self-sufficient development that was Letchworth in Hertfordshire, and would be better described as a Garden Suburb.

The estate was built at a density of 12 to the acre or better, with 70 feet between houses, a 15ft wide carriageway and 5ft wide paths. The houses

Hereford's Garden Suburb: a pair of semis

were arranged in 15 rows of five and three terraced houses and eight blocks of semi-detached houses. Every house was built of Hampton Park bricks, mortared with a mixture that includes black ash. A local design feature led to the upper storeys being clad in rough-cast, and they were roofed in red tiles from Broseley. The accommodation for all was two living rooms, three bedrooms, a kitchen, an inside WC and coal store. There was an upstairs bathroom in 16 of them (the semi-detached) and an iron bath in the scullery of others. The first keys were handed over in 1909 when the rents ranged from 4s 9d to 7s 6d a week.[46]

6 The Brickworks

There are over 300 known brickmaking sites in Herefordshire, the more significant ones are listed in the gazetteer that follows while a complete list can be found in the Herefordshire Sites and Monuments Record (www.smr.herefordshire.gov.uk). Brickworks can be divided into small, medium and large types. Maps show that there were many small brickyards, set up to build a single house and barn, which have been lost without trace leaving little more than a field name, small pond or marshy area to show their presence.

Next in the scale were brickworks owned by the large estates in the county. Originally such yards were established to provide brick for the estate's own use. Gradually however, demand came from outside the estate, both for brick and agricultural drains, and the estates were keen to satisfy this demand. The Much Cowarne estate had their brickworks at Orls Farm (HSM 31569) on the A417 which was called the Burley Gate Brickyard. It is featured on pp.115-117, and the illustration on p.32 shows some of its workers. There is a brickyard in the same parish about a mile and a half away on the A465 (HSM 32793) which is thought to be an earlier brickyard for the estate. Another estate brickworks were those of the Hampton Court estate, bought by the Arkwright family in 1810, which had four brickyards. An accounts book in the Hereford Record Office indicates the many hundreds of thousands of bricks, tiles and drains they produced in the last quarter of the 19th century. Many of these were used to build estate cottages but some were sold. We have been unable to identify the names in the book with specific brickworks on the ground, but of the four, two are described as Hampton Court Estate Brickworks—19th century (HSM 32845) and 20th century (HSM 32849)—and the other two are likely to be the ones mapped near by: Cherry Brick or Cherry Brook Orchard Brickworks (HSM 32445), and brickworks near railway tunnel (HSM 34275).

There were several intermediate sized brickworks employing a handful of people that operated independently of the large estates; the Laysters Pole brickworks being a good example of one of these (HSM 31583).

Finally there were large brickworks located near to rail links. In Hereford there were four main brickworks responsible for building the Victorian streets beyond the city walls. They operated for about a hundred years and were sometimes owned by the same people. The Albert Brickworks on the Holmer Road (HSM 32329) made a hard, red brick that didn't absorb moisture and was sold in large quantities in and out of the county. At one time it had five kilns, quarrying equipment and several long drying sheds. There is little to see there today save grass fields and a few workers cottages that have been much modernised. Just to the south of the site is the present Holmer Industrial Estate which is comprised in part of the Victoria Tile Works buildings (HSM 9414). Some of these buildings derive from the brickmaking buildings.

The Aylestone Hill Brickyard, Barr's Court (HSM 32327) in 1891 produced 600,000 bricks per year, but little is known about it. The Hampton Park Brickworks in Tupsley (HSM 31564) employed 17 people before the Second World War, had four kilns and the owner was a William Bowers who lived at Brockington House, Hafod Road, now used as council offices. He was also the builder of the Town Hall which we have featured in chapter 6. These works closed in 1937.

The Godwin brick and tile works at Lugwardine (HSM 11862) made 850,000 bricks a year in 1858. It also specialised in church tiles that were sold all over the country. They can be seen in many churches in Herefordshire including Hereford Cathedral and there are others decorating the outside of the houses in Wye Terrace, next door to the Sack Warehouse. Large brickworks were also working in Bromyard (the Linton Tile and Brickworks HSM 30573), Leominster (Barons Cross Brickworks HSM 32340 and Ebnall Brickworks HSM 38763) and Ledbury (for example the Ballard Brickworks HSM 32792).

The Hampton Park brick is extremely hard and difficult to drill. It is a brick 'second to none', heavy, strong, indestructible and with good load bearing qualities. The Holmer brick was also good, although possibly not such a high quality. In comparison the London brick, which was used in such great quantities in the county from the 1930s to the 1980s, is a soft brick, susceptible to weathering and therefore inadvisable to render, use below the damp course or in tall buildings that have to take loads.

Gazetteer of Brickworks in Herefordshire

The **Site Names** are based either on the tithe field name, a surviving name of a larger brick activity, or where both these are absent, one given by the present writers.

HSM refers to the Herefordshire Monuments and Records database number that can be sourced on the internet at http://www.smr.herefordshire.gov.uk or by a visit to the sites and monuments records office in Hereford Town Hall, where photographs are held of many of the sites.

All **grid references (GR)** in Herefordshire fall within the 100 km square 'SO', and this prefix is not shown in the Gazetteer.

OS first edition. The Ordnance Survey (OS) first produced maps for Herefordshire at a scale of 1:2500 in the 1880s. We have referred to these as 'OS first edition'.

Single use sites. Many of the records have been sourced from the Tithe Records dated between 1835 and 1845. These represent, in many cases, what we have called 'single use sites'. We do not mean to suggest that they were used just once, though they may have been, but the implication is that they had a very short life.

Clay pits. We have recorded many clay pits sourced from the OS first edition maps, and it has been only possible in a few cases to relate them to nearby brick activity. The remainder may be evidence of single use brick making, or the clay may have been used to marl the fields.

ABBEY DORE

Site Name: **Brick Close**
HSM: 32408 GR: 3830 3330
Description: This field lies on the north side of B4347 with a stream on the south-east boundary.

Site Name: **Brick Close**
HSM: 32407 GR: 3960 3460
Description: Sited on west side of Roman Road opposite to a farm. There is a stream on north boundary of field.

Site Name: **Whitfield Estate Brickworks**
HSM: 31560 GR: 4258 3380
Description: Kiln Brake is the only reference to the brickyard that was used to build Whitfield House. The kiln is below the house. The remains of another kiln can be seen at Perrystone.[47] In the Mineral Statistics of the Geological Survey of 1858 the Whitfield Brickworks are one of 25 in the county. Owned by the Rev. Archer Clive, no output is listed which suggests it was comparatively small.

ACONBURY

Site Name: **Brick Bats Meadow**
HSM: 32285 GR: 5235 3260
Description: The field is at the crossroads and there is pond extant – probably the source of the clay.

Acton Beauchamp

Site Name **Earthfield**

HSM: 32286 GR: 7030 5010

Description: Near Redmarley Farm.

Allensmore & Thruxton

Site Name: **Brick Clamp**

HSM: 32409 GR: 4420 3510

Description: Sited 700m from the B4348.

Site Name: **Upper, South, Lower Kiln Field**

HSM: 32287 GR: 4665 3570

Description: Located in a field to the south of Allensmore church.

Site Name: **Clay Pit**

HSM: 41360 GR: 4424 3205

Description: We have been unable to positively identify a brick activity in the vicinity. Sourced from the OS first edition.

Almeley

Site Name: **Brick Kilns on Batch Brook**

HSM: 38450 GR: 3291 5179

Description: Old Brick Kiln is marked on the OS first edition map about 100m east of the railway line. Where the stream has cut into the bank piles of drainage pipes can be seen today, covered in stinging nettles. No bricks could be seen. One sample collected which is a machine made pipe, probably steam or diesel driven.

Site Name: **Part of Brick Close**

HSM: 32410 GR: 3485 5343

Description: This is situated beside the A480. One of 25 brickworks listed by the geological survey of 1858. This brickworks, on land owned by a Miss Foley, the major landowner in the parish, was listed at Almeley. The source was described as clay of old red sandstone. The factory made 91,513 bricks, 171,914 draining tiles, 798 crest tiles, 2,900 tiles, and 348 squares for flooring, all per annum.

Aston Ingham

Site Name: **Kiln Pierce Field**

HSM: 32289 GR: 6852 2293

Description: This is adjacent to Newhouse Farm and probably represents a single use site.

Avenbury

Site Name: **Brick Kiln Field**

HSM: 32291 GR: 6422 5223

Description: As this is sited 300m north of Heath Farm, it was most likely for the use of the farm.

The existing pond, site of the clay pit for the brickworks, at Burgess Farm. This is a good example as to how a disused brickworks site can appear today

Site Name: **Burgess Farm Brickworks**
HSM: 31574 GR: 6515 5217
Description.: The brickworks and clay pits were shown on the 1888 edition of the OS map, but they were gone by 1964. The clay pits are now small lakes. A series of linear ditches can be seen, these may have been brick drying fields. The site lies to the east of the road which goes downhill to Bromyard. A very large pit 200m x 100m up to 2m deep, average 1.4m could be seen on a visit in 2003.

Site Name: **Brick Kiln**
HSM: 18101 GR: 6420 5220
Description: This field is to the north of Heath Farm and was probably for the use of the farm.[48]

Site Name: **Brick Close**
HSM: 32290 GR: 6540 5367
Description: This field lies between the farm and the river on a level ground and the bricks were probably for the use of the far field.

Site Name: **Clay Pit**
HSM: 41153, 41246 GR: 6612 5052, 6517 4547
Description: We have been unable to positively identify brick activity nearby.

AYLTON AND PIXLEY
Site Name: **Kiln Orchard**
HSM: 32292 GR: 6575 3751
Description: Situated 100m south of Court Farm with an existing pond, which might be the source of the clay, though it also appears to be the source of the brook flowing south-east. In the 1891 Kelly's *Directory* there is a reference to John Charles Davies, Fair Tree with Wall Hall and Pixley Farms, Ledbury, listed as a brick- and tilemaker.

Aymestrey

Site Name: **Brick Kiln Meadow**

HSM: 32384 GR: 4130 6666

Description: This is sited 600m south-east of Lower Lye between road and wood on a steep slope, and probably represents a single use site.

Site Name: **3 Clay Pits, Old Clay Pit**

HSM: 41052, 41035, 41053, 36488

GR: 4317 6082, 4315 6743, 4155 6145, 3952 6764

Description: We have been unable to positively identify a brick activity nearby for any of these clay pits. All sourced from the first edition OS map.

Bacton

Site Name **Brick Clamp**

HSM: 33646 GR: 3700 3300

Description: Members of the Archaeology Section of the Woolhope Club and members of the Monmouth Archaeological Society visited the site of Bacton Castle in 1997 where remains of a possible brick clamp were found in the woods adjoining the castle. Fragments of brick were taken away for examination.[49]

Bishops Frome

Site Name: **Bromtrees Hall Brickworks**

HSM: 32481 GR: 6375 4835

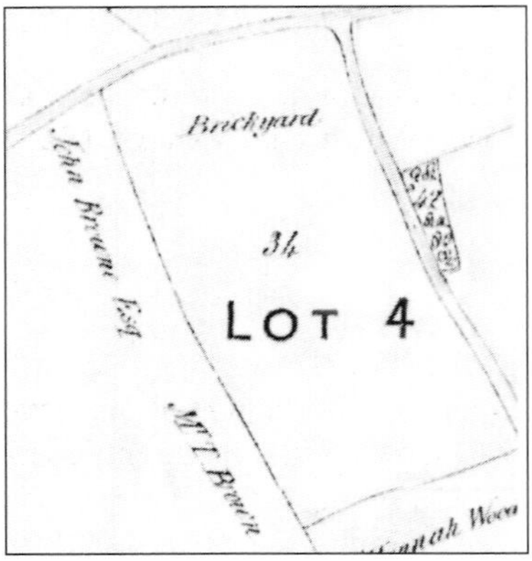

Plan of Lot 4 from a sale particulars dated 2 October 1874, showing a 'Brickyard' at Bromtrees Hall

Description: On the tithe map this field is not marked as a brickworks. Shown on first edition OS map as a small building named 'Brick Works' on the south side of the Burley Gate to Bishops Frome road. On sales particulars of Bromtrees Hall Estate dated to 2 October 1874 Lot 4 is described as a well established Brickyard with the necessary Brick Kilns, Drying Floors, Machinery and Offices requisite for carrying on a considerable trade. It is probable, therefore, that these works are the estate brickyards for Bromtrees Hall. Basil Gwilt (born in Stoke Bliss in 1923 but lived at Richley Farm from 1925 to 1955) can remember the brick ovens in the field. He thinks the cottage next to Panks Bridge was made of these bricks. The site was visited in 2003 and the only remains are the pools.

Site Name: **Smoke Acre, Ash Bed**

HSM: 32294 GR: 6680 4672

Description: This is adjacent to main road on Locks Hill, and is now partly orcharded. It probably represents a small use site.

Site Name: **Clay Pits**
HSM: 41240, 41242 GR: 6383 4853, 6530 4954
Description: Clay pit. We have been unable to positively identify a brick activity nearby for these clay pits, though see Bromtrees Hall Brickworks HSM 32481 for HSM 41240. Sourced from the first edition OS map.

Blakemere
Site Name: **Brick Meadow**
HSM: 32295 GR: 3605 4114
Description: Sited 200m west of centre of village.

Site Name: **Kites Wood Brickworks**
HSM: 31573 GR: 3566 4155
Description: Brickworks and kiln are 600m north-west of Blakemere village, and the access road (bridleway) is still in use. A brook or drain runs along the north side of the site and a small pool remains today, which is a possible source of the clay.

Site Name: **Blakemere Brickworks**
HSM: 31572 GR: 3640 4061
Description: The brickworks are 500m south-west of Blakemere Village. A building and a kiln are shown on the OS first edition. There is only a pond remaining today.

Bockleton
Site Name: **Brick Kiln at Quinton**
HSM: 31584 GR: 5786 6163
Description: Brick Kiln is 350m east-south-east of Lower Quinton Farm. There are associated quarries including one which was flooded in 1890. The site is situated 900m from the county boundary.

Bolstone
Site Name: **Brick Piece**
HSM: 32296 GR: 5495 3289
Description: Tithe field numbers 35, 36, and 37 all lie on the north side of the minor road to the west of Bolstone Court Farm. 35 is Brick Piece, 26 is the Rough Ground and 37 is Lime Kiln Field.

Bosbury
Site Name: **Brick Kiln/Brick Clamp Orchard**
HSM: 18324/32411 GR: 6860 4270
Description: This site is close to B4214, and about 1km from village, and north-east of Cold Green. As there are two names for the same site, we take it to mean that it was a very suitable place for making bricks.

Site Name: **Brick Kiln**
HSM: 18279 GR: 7130 4400
Description: Brick Clamp Piece is situated north of Slatch Wood.

Site Name: **Brick Clamp Piece**
HSM: 32412 GR: 7178 4400
Description: This is situated beside minor road, and 2km from village centre.

Site Name: **3 Clay Pits**
HSM: 41245, 41247, 41291 GR: 6860 4604, 6873 4513, 7029 4538
Description: We have been unable to positively identify a brick activity nearby. Sourced from the first edition OS map.

Brampton Abbotts

Site Name: **Orchard by the house and Brickyard**
HSM: 32413 GR: 6060 2690
Description: This probably represents a single use site and is situated 500m from village centre. There is a pool in the field which is a probable source of the clay.

Site Name: **Brickyard**
HSM: 18157 GR: 6051 2785
Description: This is sited north of Brickyard Cottages.

Site Name **Clay Pit**
HSM: 41293 GR: 5895 2260
Description: We have been unable to identify a brick activity nearby. Sourced from the OS first edition map.

Brampton Bryan

Site Name: **Stables and Brickyard**
HSM: 32414 GR: 3690 7260
Description: Situated on the western edge of the village.

Site Name: **Clay Pit**
HSM: 41019 GR: 3773 7168
Description: We have been unable to positively identify a brick activity nearby. Sourced from the OS first edition map.

Breinton

Site Name: **Brick Kiln Meadow**
HSM: 23163 GR: 4770 3905
Description: No features noted.

Site Name: **Brick Kiln**
HSM: 32297 GR: 4750 4020

Description: On south side of the Breinton moraine (in fact an esker) and facing the Wye. It is immediately north-west of Little Breinton.

Site Name: **Upper and Part of Brick Kiln Meadow**
HSM: 32415 GR: 4780 3900
Description: Beside the River Wye and 3km from Wye Bridge, Hereford.

Site Name: **Brick Kiln**
HSM: 23046 GR: 4756 4020
Description: Brick Kiln north of Little Breinton.

Brimfield

Site Name: **Brick Kiln, Pasture in Brick Kiln**
HSM: 32299 GR: 5230 6780
Description: Brimfield Brook runs along north boundary of the field which is 500m west of the centre of Brimfield.

Site Name: **Brickyard Piece, Brickyard**
HSM: 32298 GR: 5080 6760
Description: Beside the Leominster Canal, some distance from nearest buildings (200m) or road (400m) so was probably used for the canal itself, or possibly as a source of bricks for sale which could be distributed along the canal.

Site Name: **Clay Pit**
HSM: 41036 GR: 5205 6632
Description: We have been unable to positively identify a brick activity nearby. Sourced from the first edition OS map.

Brinsop & Wormsley

Site Name: **Brick House Field Orchard**
HSM: 32416 GR: 4480 4610
Description: Located 300m above Brinsop Court.

Brobury with Monnington on Wye

Site Name: **Clay Pit**
HSM: 41257 GR: 3490 4490
Description: We have been unable to positively identify a brick activity nearby. Sourced from the first edition OS map.

Bromyard & Winslow

Site Name: **Brick Clamp**
HSM: 32300 GR: 6460 5485
Description: The site is presently occupied by the school and 300m north-west of the centre of Bromyard. This may have been the brick and tile works listed in Kelly's *Directory* of 1891 owned by Rt. Lewis in Sherford Street.

Site Name: **Brick Piece/Brick Meadow**
HSM: 32472 GR: 6210 5410
Description: The only information comes from the tithe map.

Site Name: **Sheep Street Brickyard**
HSM: 32284 GR: 6540 5450
Description: A deed to a parcel of land in Hereford Record Office (W34/7) refers to a brickworks. The Conveyance of 1 May 1902 states: 'Parcel of land on south side of Sheep Street, 1469 sq. yards and a parcel of land or Brickyard on south side of property having a frontage on New Road 2467 sq. yards with all buildings, kilns, sheds and clay mill thereon.' The property was formerly part of Collington Croft.

Site Name: **Streamhall Brickworks**
HSM: 34697 GR: 6500 5400
Description: 'Bricks fused together found in my garden (Whitbourne) have the stamp Streamhall, Bromyard.'[50] See also Linton Tile and Brickworks HSM 30573.

Site Name: **Hackley Farm Brickworks, Winslow**
HSM: 11196 GR: 6329 5328
Description: Referred to as Old Kiln on west bank of Hackley Brook. The structures have gone now, and it was referred to as 'Old Brick Field' because of debris lying over surface. On the first edition OS map a kiln and clay pits are recorded in a 6ha field bounded by Hackley Brook on one side and adjacent to the road to Bromyard. This was a larger brickworks. Slight remains of foundations of kiln could be seen by the brook in 2003. A sample brick found.

Site Name: **Linton Tile and Brickworks (Stream Hall Quarry), Bromyard**
HSM: 30573 GR: 6680 5415
Description: This large 19th-century brickworks was adjacent to the Bromyard to Worcester road with clay pit about of 3.9ha behind it shown on the 1964 OS map. The Worcester to Bromyard railway ran alongside the works and they had their own sidings. A much smaller quarry is shown in the same place on the first edition map, and the brick works are shown on this edition with a smaller brickworks to the north of the road. Neither of these brickworks are shown on the 1841 tithe map.

Stream Hall in 1842 was a smallholding with house and farm buildings owned and occupied by John Fudger. By 1878 it was a sandstone quarry employing many Bromyard people and was owned by William Finney who lived on Bromyard Downs. The following year the firm was making bricks and tiles and the Linton Tileries was called Stream Hall Quarry. William Finney's firm was making bricks and tiles, and it became a limited company in the 20th century.

They supplied tiles for the Admiralty and War Office in 1914. The works secured the largest contract for paving tiles ever placed in the country by a foreign government for the Egyptian State Railways which were made from Old Red Sandstone marls.[51]

The railway to Yearsett (3.5 miles from Bromyard) was opened in 1874 and transported the bricks and tiles. In 1887 the line was extended to Bromyard and Leominster and there were sidings at the tile works.

R. Lewis was listed as owner/worker of Hodgebatch, Linton, in the Kelly's 1891 *Directory*. In 1968 the works employed about 50 men when it was called a tile works, so presumably it no longer made bricks. The tile works closed in the 1970s, though small scale work continued until the 1980s. Landscaping occurred in the same decade and by 2000 was being used as a Council waste site.

General view of the Linton Tile and Brickworks with, top, workers at the clay face

Site Name: **8 Clay Pits**
HSM: 41151, 41105, 41145, 41106, 41154, 41152, 41148, 41147
GR: 6339 5451, 6375 5674, 6289 5400, 6297 5727, 6685 5385, 6374 5439, 6331 5343, 6325 5351
Description: We have been unable to positively identify brick activity nearby for any of these clay pits, though they could have been used by any of the brickworks in Bromyard. Sourced from the first edition OS map.

Buckton and Coxall
Site Name: **Brick Kiln**
HSM: 18189 GR: 3750 7328
Description: This was situated north-east of Parsons Pole Bridge.

Burghill
Site Name: **Brick Kiln**
HSM: 32301 GR: 4847 4543
Description: The site is now occupied by St. Donat's Farm beside the present A4110. It may have been the brickworks listed in the 1858 mineral statistics of the Geological Society. The land was owned by John Barrel Fencot and the manufacturer was James Nicholls. Brick and tiles were made. On the 1885 map the Brick and Tile Works are shown on the south side of the road, about 100m to the south-east of Brick Kiln Farm (now St. Donat's Farm).

Site Name: **Brickhouse, Perry Gobbitt**
HSM: 32418 GR: 4870 4640
Description: This is situated 800m north-east of St. Donat's Farm. See Brick Kiln (HSM 32301).

Site Name: **Brickyard**
HSM: 32417 GR: 4610 4580
Description: Adjacent to Broomhill Farm at Tillington Common and 200m north-west of Brickhouse Farm.

Burrington
Site Name: **Brick Kiln Piece**
HSM: 32419 GR: 4790 7340
Description: High on steep slope.

Site Name: **Brick Barn Farm**
HSM: 31587 GR: 4365 7140
Description:
Brick Barn Farm is 750m south-south-east of Burrington Bridge. It is so called either because it is a brick building in a stone country or because it is in an area of a kiln. A visit in 2003 revealed no sign of a brickworks.

Byford

Site Name: **Portway Brickworks**

HSM: 31571 GR: 3846 4418

Description: The brickworks are in rough ground 300m on the Hereford side of the Portway Inn on the A438. It lies to the north of the road by Maddle Brook and is on land owned by the Garnons Estate, and may therefore have been the estate brickworks. Today there are no buildings, but there is one large and one small pond, lightly wooded. Visited in 2003.

Canon Frome

Site Name: **Brick Close**

HSM: 18451 GR: 6630 4315

Description: Located east of Childer Wood. A visit in 2003 revealed no remains of brickworks.

Site Name: **Clay Pit**

HSM: 41282 GR: 6542 4317

Description: We have been unable to positively identify a brick activity nearby, though see 18451. Sourced from the first edition OS map.

Canon Pyon

Site Name: **Old Brick Kiln**

HSM: 32302 GR: 4536 4975

Description: This lies on the north-facing slopes of Pyon Hill.

Site Name: **Ovens Hill**

HSM: 32303 GR: 4785 5060

Description: In this area are Brickyard Cottage, fields nos. 524 Coal Pits, 532 New Cinders. Ovens Hill could refer to charcoal burning, especially in reference to 524 and 532, since there is plenty of woodland on the slopes to the west. In relation to the cottage however, it is more likely to be brick ovens.

Site Name: **Upper Derndale Brickyard**

HSM: 36519 GR: 4705 4856

Description: 'I was born at Canon Pyon. At Upper Derndale, the pool there was supposed to be for brick making. I was told that when in the 1850s the farms were refurbished they made the bricks on the premises to build the front wings on the house'.[52]

Castle Frome

Site Name: **Brick Clamp**

HSM: 18123 GR: 6765 4585

Description: South-west of Buckwood Farm.

Clifford

Site Name: **Little Brick Close**
HSM: 19250 GR: 2460 4545
Description: Field name 'Little Brick Close' is recorded on the tithe map 1842.

Site Name: **Great Brick Close**
HSM: 19252 GR: 2485 4545
Description: Field name 'Great Brick Close' is recorded on the tithe map 1842.

Site Name: **Middle Brick Close**
HSM: 32304 GR: 2470 4550
Description: Lies on outskirts of the village with two lanes for access. Later the railway was built beside the site. Field 235, Little Brick Close HSM 19250 is to south-west and 242 Great Brick Close HSM 19252 is to north-east.

Site Name: **Brick Close**
HSM: 32423 GR: 2640 4520
Description: Sited 100m south-west of the Farm. Note 'Pits' on opposite side of road.

Site Name: **Bricklands**
HSM: 30485 GR: 2420 4288
Description: The house called Bricklands is at the Moors, on old road sunken way (HSM 30484) within a parkland (HSM 21741) landscape.

Site Name: **3 Clay Pits**
HSM: 41164, 41163, 41162 GR: 2988 4522, 2574 4618, 2549 4601
Description: We have been unable to positively identify brick activity nearby, though 41163 and 41162 could have been used by Middle Brick Close (HSM 32304). Sourced from the first edition OS map.

Coddington

Site Name: **Brick Kiln Orchard**
HSM: 32427 GR: 7210 4350
Description: Now in a wooded area.

Site Name: **Clay Pit**
HSM: 41286 GR: 7190 4837
Description: We were unable to positively identify this with brick activity nearby. Sourced from the first edition OS map.

Collington

Site Name: **Kiln Meadow**
HSM: 32306 GR: 6552 6020

Description: This site is on the flat and towards a valley bottom, suggesting that it is a brick kiln, though use for lime should not be ruled out. It lies 300m south-west of Brick Kiln Meadow HSM 32305 which is tithe field no 89.

Site Name: **Brick Kiln Meadow**
HSM: 32305 GR: 6574 6042
Description: This site probably represents a single use and it lies 800m east of Collington on fairly steep slope. It lies 300m north-east of Kiln Meadow HSM 32306 which is tithe field no 137.

Site Name: **Brick Kiln Meadow**
HSM: 32420 GR: 6505 6005
Description: Situated opposite the chapel beside B4214, it is probably another single use site and is tithe no 137. It lies 300m south-west of Kiln Meadow 32306.

Site Name: **Clay Pit**
HSM: 41110 GR: 6427 5933
Description: We were unable to positively identify this with brick activity nearby, though may well have been used by Kiln Meadow HSM 32206 or Brick Kiln Meadow HSM 32420. Sourced from the OS first edition map.

COLWALL

Site Name: **Colwall brick and tile works**
HSM: 31561 GR: 7483 4315
Description: The works on the site are 1km north-west of Colwall Stone and 400m north of the nearest road. There was one large building, two kilns and one outbuilding, and the clay pit was a pool even in 1884. Today's site is wooded with pool extant. Part of a substantial brick wall remains in the garden of owner (in 2003). The bricks are reputedly extremely hard, and it is said that most of Colwall is built of them.

It was run by the Ballard family (see Ledbury) and the occasional brick is stamped Ballard. Decorative bricks like those on the house down the lane, 'Labour's Rest', where the workers were housed, and curb stones were also made.

'I think the works closed because the clay needed a lot of drying and burning and eventually couldn't compete with clays that needed less fuel. The story is that one day the workmen turned up for work but the manager didn't and that was it. Marl Cottage was the old site office and I converted it to a house. We have found small trolleys in the pool which is very deep. Only the remains of a pump house stands today and a curiously built oak shed, made on the Ballard Estate'.[53] The size of the pool indicates that over 4.5 million bricks were made, enough for about 200 small houses.

'Fred Ballard, the eldest son of Stephen Ballard was the brickmaker. His sister Ada designed decorative bricks that were made at the works. The works were quite short lived, Fred never stuck at anything for long.'[54]

'The Ballard family had numerous business enterprises in Colwall, the brickworks was one of them, begun by Fred. He learnt brickmaking at Hamletts in West Bromwich and he began the brickworks in Colwall around 1890. The clay was moved by steam driven machinery into a pugmill, the fuel used was gas from the Colwall Gasworks. The bricks were hand pressed, there were three kilns, two longitudinal fired and one round one. There were short square brick chimneys. Three workers cottages, a boarding house and an office were built for the staff, some of whom came from West Bromwich. The bricks were similar to Ruabon bricks being very hard with a smooth finish. The yard was closed soon after 1900. The bricks were too expensive compared to Fletton bricks due to the large amount of coal used to fire them. The kilns were dismantled by government order during the Second World War and the heavy machinery was melted down. I estimate about 2.5 million bricks were made here.'[55]

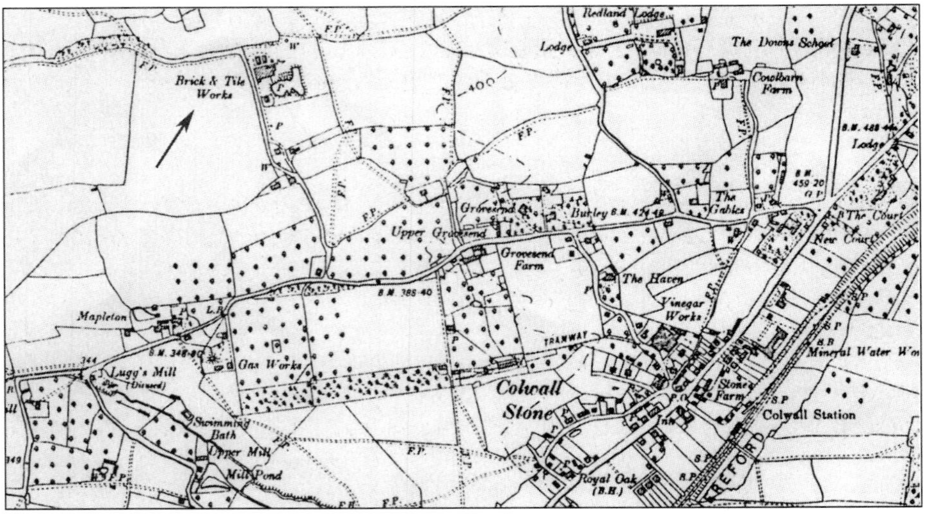

The Colwall Brickworks: Top left: Existing wall of the brickworks in owner's garden; Top right: Decorative bricks from the works used in Labour's Rest; Bottom: Map showing the location of the brickworks arrowed (Hereford 36NW)

Cradley

Site Name: **Brickclose**

HSM: 32426 GR: 7270 4830

Description: This can be found next to Beanhouse Farm.

Croft and Yarpole

Site Name: **19th-century brick kiln, Croft Castle**

HSM: 31910 GR: 4590 6532

Description: A levelled area was observed with a large amount of broken brick, which was interpreted as a brick kiln, and dated to the late 19th century on the typology of the bricks. A track leads west from the brick kiln towards Croft Castle about 600m to the east. Presumably this was a brick kiln used to make bricks for one of the phases of Croft Castle. It lies about 200m to the south-east of the earlier 17th-century brick kiln (HSM 31909).[56]

Site Name: **17th-century brick kiln, Croft Castle**

HSM: 31909 GR: 4571 6552

Description: A levelled area was observed with a large amount of broken brick and tile in a semi-hidden pile below a tree stump. This was interpreted as a brick kiln, dated to the late 17th century on the typology of the bricks. It is located around a small pond shown on the first edition OS map, and a track leads west from the brick kiln towards Croft Castle. The kiln is about 475m east of the castle and presumably was a brick kiln used to make bricks for one of the phases of the house. It lies about 200m to the north-west of the later 19th-century brick kiln (HSM 31910).[57]

Site Name: **Breck Kilns**

HSM: 18121 GR: 4557 6467

Description: 'Breck Kilns' is the name of the field, and assumed to be a local variation on 'Brick Kilns'. It is situated north of Grove Barn and is not shown on the OS 1964 edition.

Site Name: **Clay Pit**

HSM: 41030 GR: 4709 6573

Description: We have been unable to positively identify a brick activity nearby. It is shown on the first edition OS map.

Dilwyn

Site Name: **Brick Meadow**

HSM: 32309 GR: 4183 5478

Description: This is a field on the north-west outskirts of village beside the road to Dilwyn Common.

Site Name: **Brick Orchard**

HSM: 32310 GR: 3850 5396

Description: It lies in the fork of a minor road and farm drive. We assume it was a single use site.

Site Name: **Brick Lane Hopyard**

HSM: 32307 GR: 4105 5606

Description: Now an orchard, it lies 100m west of Bidney Farm. No doubt a single use site.

Site Name: **Brick Kilns**

HSM: 32308 GR: 4097 5340

Description: This is a large site with no extant route to roads or The Homme. It lies 550m north-west of The Homme.

Dinedor

Site Name: **Brick Field**

HSM: 32429 GR: 5420 3710

Description: Near the B4399 as it climbs away from river, it is sited between the road and later railway.

Site Name: **Kiln Field**

HSM: 32311 GR: 5375 3686

Description: The type of kiln is not stated and may be for lime. It is sited 200m east of the village.

Dinmore

Site Name: **Clay Pit**

HSM: 41086 GR: 4800 5062

Description: We have been unable to positively identify a brick activity nearby. It is shown on the first edition OS map.

Docklow & Hampton Wafre

Site Name: **Brickyard Plantation, Buckland**

HSM: 31435 GR: 5535 5645

Description: Brickyard Plantation is a wood containing lakes about 300m south of Buckland. 'The clay was dug from the lake pits to the south-west and moved to the house by railway. Parts of the railway still exist, incorporated into the farm buildings'.[58]

Donnington

Site Name: **The Brick Clamp**

HSM: 32428 GR: 7100 3400

Description: It lies in the centre of three farms and could have served any or all.

Dorstone

Site Name: **Brick Close**

HSM: 32424 GR: 3107 4110

Description: This is sited 1km south of the village, 150m along track from Pit Road, and on wooded slopes.

Downton

Site Name: **Piece by Brick Barn in Brick Barn Leasow**
HSM: 32421 GR: 4230 7210
Description: A field name taken from the tithe map and sited below Nacklestone Farm.

Site Name: **Downton Castle Brickwork**
HSM: 31588 GR: 4480 7565
Description: These are probably the estate brickworks and are located 750m north-east of Downton Castle. A very light brick scatter was observed during a site visit in 2003. There is a causeway through the wood to public road. Sample of brick and land drain was found. The site was present in 1887 and 1904 and probably active on both dates. A later visit found diggings that were clearly carried out by tramway or cartway. Parts of the shed and kiln are still there and a stockpile of curved or decorative bricks. All the diggings remain open, apparently exactly as they were abandoned early in the century.[59]

Site Name: **Brick Kiln Field, Old Church**
HSM: 18151 GR: 4375 7432
Description: Probably for the use of the farm.

Eardisland

Site Name: **Brickfield Plantation**
HSM: 31579 GR: 4230 5760
Description: Brickfield Plantation with a quarry and lake was shown on the first edition of the OS map and as field 240 on the tithe map. This is probably the brickfield mentioned with 24 other brickworks in Herefordshire on the Geological Survey of 1858 as being at Burton Court, using alluvial clay, owned by the Rev. W. Evans. Evans was a large landowner in 1842 and though he didn't own field 240 he owned the fields surrounding. Possibly 16 years later it had spread into his land, or he had bought it. Much of the area had been built up, only a pond and bank in view in 2003.

Eardisley

Site Name: **Eardisley Brickworks**
HSM: 31565 GR: 3160 4965
Description: This brickworks had its own claypit and lies 500m east-north-east of Eardisley village. The site is not shown on the OS Explorer 201 map but the track north to the Almeley road exists by the field boundary. No trace remained when visited in 2003. The old tramway from Hay-on-Wye to Kington runs 200m to the west.

Site Name: **Brick Kiln**
HSM: 19174 GR: 2975 4865
Description: On the tithe map the fields are called Brick Kiln Piece and Kiln Field and are situated north-west of Eardisley Park.

Site Name: **Brick Kiln**
HSM: 19152 GR: 2970 5200
Description: Three more fields: Brick Kiln Croft, Brick Kiln Meadow, Brick Kiln Wood, and are situated north-west of Little Quebb.

Site Name: **Brick Kiln Croft/Meadow**
HSM: 32425 GR: 2970 5190
Description: Situated 1km from the A4110 above (and to the north of) the village.

Site Name: **Brick Orchard**
HSM: 32314 GR: 3107 5282
Description: Sited at Lower Spond, now destroyed, and probably used to build (part of) the farm.

Site Name: **Clay Pit**
HSM: 41167 GR: 3161 4957
Description: We have been unable to positively identify brick activity nearby, though see Eardisley Brickworks HSM 31565. Sourced from the OS first edition map.

EASTNOR
Site Name: **Brickyard Orchard**
HSM: 32315 GR: 7315 3780
Description: Sited by the A 438, and still an orchard today behind Woodside Buildings.

EATON BISHOP
Site Name: **Brick Close**
HSM: 32430 GR: 4390 3765
Description: Located on the opposite side of the road to the Common and next to field 14 on the tithe map. Commons were often sites of brick activity and maybe the brick workers had an interest there also.

Site Name: **Brick Close**
HSM: 32316 GR: 4390 3765
Description: Sited on Littlemarsh Common beside the road from Crossway Farm.

Site Name: **Brick Close**
HSM: 32317 GR: 4407 3911
Description: Situated 200m west of the church on the hill top and there is a pool to east of the site. 'These were stables at one time, an isolation unit for the stud farm that had stables throughout the village. The buildings were demolished and the bricks used in buildings close to the church.'[60]

EDVIN LOACH AND SALTMARSHE
Site Name: **Brick Kiln Meadow**
HSM: 32318 GR: 6692 5911
Description: The area today is covered by large ponds.

Site Name: **Clay Pit**
HSM: 41114 GR: 6597 5865
Description: We have been unable to positively identify a brick activity nearby. Sourced from the OS first edition map.

Edwin Ralph

Site Name: **The Brick House Homestead**
HSM: 32319 GR: 6480 5792
Description: Now Brick House Farm. No details of Brick Kilns and it may be so called because a brick house in the area was unusual.

Site Name: **Clay Pit**
HSM: 41107 GR: 6364 5798
Description: We have been unable to positively identify a brick activity nearby, and identified from the first edition OS map.

Eggleton

Site Name: **Clay Pits**
HSM: 41241 GR: 6257 4522
Description: We have been unable to positively identify a brick activity nearby, and identified from the first edition OS map.

Elton

Site Name: **Brick Kiln**
HSM: 11834 GR: 4520 7138
Description: Sited north-west of Elton Farm.

Ewyas Harold

Site Name: **Pontrilas Brickworks**
HSM: 11864 GR: 3940 2750
Description: This large brickworks stood near the junction of the A465 and the Prontrilas-Rowlestone road from the mid-19th century until the First World War. A petrol filling station and other businesses now occupy the site, but the adjacent row of red brick houses, known as Brickyard Row, remains.[61] It was owned at one stage by the Hampton Park Brick and Tile Co. of Hereford.

Site Name: **Brick Piece**
HSM: 32320 GR: 3906 2690
Description: This may have been a satellite of main site by the main road since it was next to a farm and only 600m south-west of Pontrilas Brickworks (HSM 11864).

Site Name: **Clay Pit**
HSM: 41350 GR: 3936 2727
Description: Although we have been unable to positively identify this with a brick activity nearby, it could well have been a source of clay for the Pontrilas Brickworks HSM 11864. It is shown on the first edition OS map.

Eye, Moreton and Ashton

Site Name: **Brick Buildings**
HSM: 31585 GR: 5009 6263
Description: A farm named Brick Buildings isolated on the flood plain of the Main Ditch. A possible indicator of brick kilns locally, but no clay pits were shown on the 1894 map.

Site Name: **Brick & Barn Meadow**
HSM: 32422 GR: 5085 6395
Description: Situated 200m north-west of Berrington Hall.

Fownhope

Site Name: **Brick Meadow**
HSM: 32321 GR: 5835 3533
Description: This large field is on clay soil and sited below woods. Woods indicate a source of fuel.

Foy

Site Name: **Brick Field**
HSM: 32432 GR: 6005 2820
Description: Here is an extensive site beside the River Wye. The river may have been used to transport the products away from the works.

Site Name: **Brick End**
HSM: 31553 GR: 6017 2887
Description: The OS Explorer 189 Brick End refers to a group of houses, and the associated one or two fields. These fields are now arable, and were probably the site of the works and pits. Although these are no positive remains there is one brick building at the end of the site which is used as a garage for 'White Walls'. It is out of context with this building (rendered) and others in vicinity (stone). It is therefore likely to be part of a brickworks.

Site Name: **Brick Field**
HSM: 32431 GR: 6010 2865
Description: Another extensive site in this parish and adjacent to Brick End HSM 31553.

Garway

Site Name: **Brickyard Field**

HSM: 32322 GR: 4740 2197

Description: On the tithe map Lower Brickyard Field is no. 705, and is beside the road and Coed Lank Wood.

Grafton

Site Name: **Nichols' Brickworks**

HSM: 32790 GR: 5000 3600

Description: A brickworks was described in the Geological Memoirs of 1858. The manufacturer was J. Nichols, and the freeholder was J.W. Jones. Two brickworks (out of 25 in Herefordshire) were listed in Grafton on the 1858 Geological Survey. They both used clay formed on the Red Sandstone'. Nichols' works made common bricks, about 200,000 p.a. The location is not known. It may have been near to the Hudson Brickworks (see HSM 31557), which was the other manufacturer mentioned working at the same time, but this is only speculation.

Site Name: **Brick Kiln Inn (now Grafton Inn). Hudson's Brickworks**

HSM: 31557 GR: 5023 3618

Description: The inn is named after the kiln, and the 2.6ha field behind and to the north is labelled as having clay pits in 1887, and it is assumed that they were all out of use at this time. Two brickworks (out of 25 in Herefordshire) were listed in Grafton on the 1858 Geological Survey. They both used clay formed on the 'Red Sandstone'. This manufacturer was George Hudson on land owned by R. Minton. He made an unspecified number of bricks and drains.

A three-storey building at the Grafton Inn, possibly once an office for the brickworks

Site Name: **Clay Pit**

HSM: 41370 GR : 5028 3632

Description: Grafton has had more than its fair share of brickworks, so this pit could have been used by any or all of them. It is however, particularly close to Nichols' Brickworks, HSM 32790. It was sourced from the OS first edition.

Grendon Bishop

Site Name: **Brick Kiln**

HSM: 18475 GR: 6070 5740

Description: Brick Kiln Field is situated east of Batchley Farm.

Hampton Bishop

Site Name: **Brick Ground**
HSM: 32434 GR: 5560 3740
Description: Situated 400m south of the B4224, and in open country today.

Site Name: **Clay Pit**
HSM: 41368 GR: 5446 3895
Description: This may be associated with Brick Ground, HSM 32434. Sourced from the OS first edition map.

Site Name: **Clay Pit**
HSM: 41364 GR: 5273 3975
Description: We have been unable to positively identify a brick activity close by. Sourced from the OS first edition map.

Hampton Charles

Site Name: **Old Brick Works**
HSM: 22485 GR: 6018 6082
Description: Abbott Thomas of The Stone, Hampton Charles, Tenbury, is listed as a brick- and tilemaker in the 1891 Kelly's *Directory*. Details of the brickfield were included in sale particulars for the sale of Bockleton Estate, 5 July 1905.

Hatfield & Newhampton

Site Name: **Brickland Field**
HSM: 32312 GR: 5967 5862
Description: Now in a large field. This is in Hatfield and Newhampton now, but was in detached part of Docklow at the time of the tithe apportionment.

Haywood, Treville & Newlands

Site Name: **Brick Close Orchard/Meadow**
HSM: 32436 GR: 4890 3660
Description: Now a partly orcharded site, with a spring in the south-east corner.

Site Name: **Brickyard**
HSM: 32324 GR: 4272 3130
Description: A large field in common land.

Hentland & Harewood

Site Name: **Brickfield Orchard**
HSM: 32326 GR: 5295 2839
Description: A large field to the north of Harewood Park Farm, and possibly used only for the farm.

Site Name: **Brick Close**
HSM: 32325 GR: 5418 2438
Description: This is a large field beside A4137 in Hentland. There is a pond in the adjacent field to the south which may have been the claypit that provided the clay.

Site Name: **Miles Field & Brick Meadow**
HSM: 32435 GR: 5365 2780
Description: This one is sited in open country by a wood. There are no tracks, so it is likely to have been a single use site.

HEREFORD

Site Name: **Brickyard**
HSM: 32323 GR: 5268 4021
Description: This site is one of many smaller sites in the City of Hereford and is on the south-east corner of Ledbury Road and Church Street. Hampton Park Brick and Tile Works (HSM 31564), is only 300m south-east of this site.

Site Name: **Old Brick Yard**
HSM: 32433 GR: 5230 4020
Description: This is now covered with the houses of Hafod Road.

Site Name: **Stonebow Brickyard**
HSM: 32282 GR: 5170 4000
Description: We have sourced this from the records in Hereford Record Office (B43/25-7, References and Accounts) which shows bricks supplied to Richard Pritchard builder, patent brick- and tilemaker at Stonebow, Hereford in 1853. One record shows that on the 18 January 500 hard bricks cost 12s 6d, which were taken to the house at Marden of Mr. James Good of Upper Lyde.

Site Name: **Hampton Park Brick and Tile Works**
HSM: 31564 GR: 5275 3985
Description: A large brick and tile works 2km from Hereford city centre. There were adjacent clay pits, up to 5 drying sheds and 4 kilns which were extant in 1886. It is now surrounded by housing and the site is a recreation ground and nature reserve.

'I worked at the Hampton Park brickworks between June 1936 and October 1937, while waiting to work on the Post Office. I used to run the bricks down from the drying shed to the kilns. There were 17 workers including the time-keepers, and two lorries. Two worked in the quarry, two in the entrance shed. The owner was William Bowers who lived at Brockington House, Hafod Road, Hereford. Bricks from here built the Odeon Cinema and the Ritz in Commercial Road. Common and pacing bricks were made. There was one long kiln and two round ones. The big kiln was for the common bricks. Bricks were machine made. There was a big shed, bricks were stacked on separate boards and steam from the machinery helped to dry them in the 'green shed' prior to taking up to the kiln. Most of the people there were elderly chaps, so I don't know if any would be surviving now.'[62]

A sales note of 1887 shows Hampton Park Brickworks was owned by G.J. Wilson. The 1891 Kelly's *Directory* lists the owner as George John Wilson of Hereford and Pontrilas R S O. Hereford Brick and Tile Co. Ltd., 20 East Street listed in Kelly's 1917

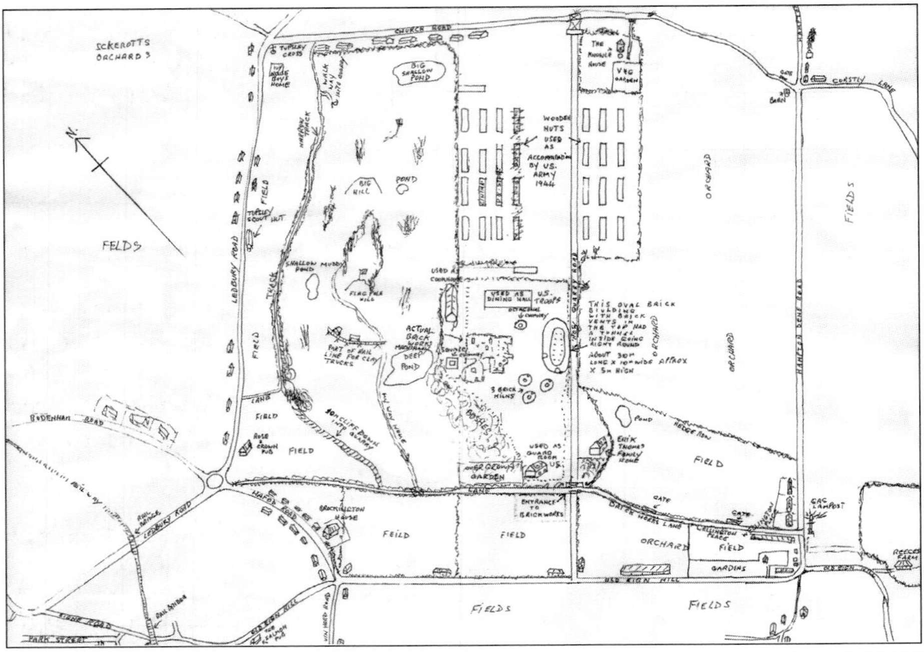

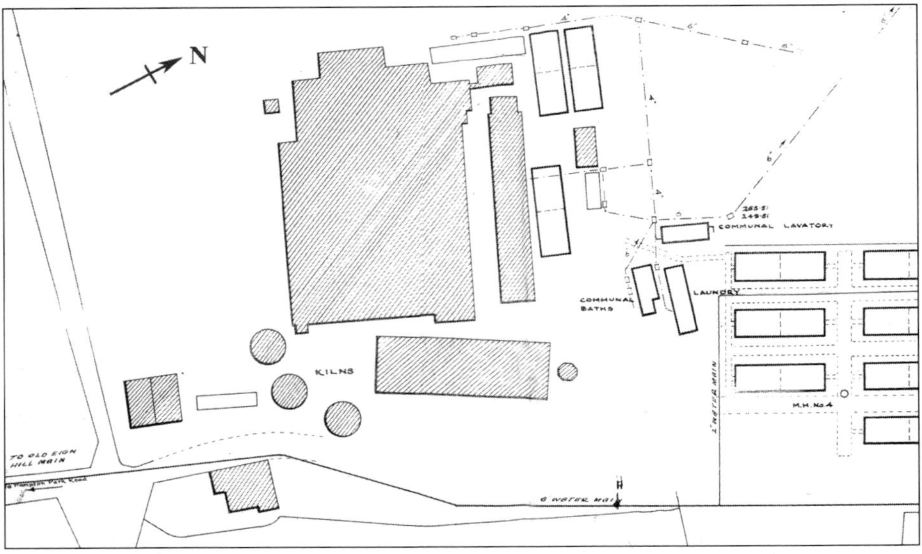

The sketch plan at the top was produced by Mr. S D Barrow and shows the site of the Hampton Park Brickworks at the end of the Second World War. The lower plan shows the site in 1947, shortly before demolition began

Directory as the Hampton Park Works. Founded as a small concern by Mr. Wilson and in 1891 taken over by Mr. W.E. Britten's Hereford Brick and Tile company.

It had continuous kilns, not extinguished once lit in 1892 until 1915 and made 5,000,000 bricks a year. It was closed in the late 1930s, just before the Second World War. The bricks were cherry red. Used as an American Army Camp in the Second World War until D Day when the military all left. It is now a local nature reserve. A commemorative brick at the base of Kington Hospital 1887 to mark Victoria's Jubilee indicates bricks went as far as Kington.[63]

Site Name: **Brick Kiln, Aylestone Hill**
HSM: 26973 GR: 5200 4000
Description: A small brick kiln, probably that mentioned in 17th-century archives, was discovered at the residence of Mr. G. Gobat on Aylestone Hill.

'I came to the conclusion that it was a brick kiln fired from the north side where the ground is at a lower level. Made of bricks with a domed top, part of which had broken away. It had filled with rubbish.'

In the Archaeological Report for 1946 is the item, 'well at Hampton Dene' and a mention of a brick kiln on Aylestone Hill is made. 'I suggest that this is the kiln mentioned in the archives of the 17th century as the clay in the vicinity is excellent and suitable for brick making.'[64]

On the Tupsley tithe map of 1842, field number 53 is called 'Brick Kiln, Sheds and Land'. This may have been the brickworks mentioned by the Geological Survey of 1858. It was called Barr's Court Brickworks, using mixed red and yellow clay, near or in Hereford. The owner was Mrs. Poole, the manufacturer was B. Pritchard. Common brick and pipes were made, 500,000 per annum. See also HSM 32327, Aylestone Hill Brickyard.

Site Name: **Aylestone Hill Brickyard**
HSM: 32327 GR: 5160 4090
Description: Now covered by Barrs Court houses, and possibly used to build them. Do not confuse this one with Brick Kiln, HSM 26973. It may be the brickyard described in the 1858 Geological Survey also as Barrs Court. The material was Red Marl, the owner was Mrs. Poole and the manufacturer was E.L. Bristow. Bricks and pipes were made, 600,000 and 200,000 respectively per annum. Possibly the brickworks were later owned by Ralph, Preece, Davies and Co. listed in the 1891 Kelly's *Directory* as being at Barrs Court Station. See also The Albert Steam Pipe and Brickyard HSM 32329 at Holmer.

Site Name: **Part of Brickfield & Pasture**
HSM: 32437 GR: 5190 5035
Description: This site is just outside the present extent of Hereford housing.

Site Name: **Brick Piece**
HSM: 32328 GR: 5005 3855
Description: This site has completely vanished under present day housing, and lies between the present A465 and the later tramroad (now railway).

Site Name: **3 Clay Pits**
HSM: 41363, 41365, 41366 GR: 5280 3995, 5268 3980, 5265 3964

Description: Although we cannot be sure about its connection with brick activity, 41363 is not far from Brickyard, HSM 32323. We have been unable to positively identify a brick activity nearby, though any local Hereford brickworks could have used the second and third pits. Sourced from the OS first edition map.

Holme Lacy

Site Name: **Brickworks**
HSM: 33848 GR: 5582 3470

Description: Narrow, small bricks were dug out of an arable field over an area about 0.5ha. These bricks disintegrated into nothing when exposed. The site was interpreted as a brick pit with unfired bricks. Possibly this kiln was the one that made the first (16th century) Holme Lacy House bricks.

Site Name: **Brick Kiln, Brick Kiln Wood**
HSM: 18546 GR: 5480 3480

Description: In Brick Kiln Meadow tithe field number 190.

Holmer

Site Name: **The Albert Steam Pipe and Brickyard**
HSM: 32329 GR: 5070 4221

Description: 'Messrs Ralph, Preece, Davies and Co, Proprietors, (of this) important works, have been established for about 50 years. Situated close to the joint lines of the Great Western and North Western Railway and with a fine frontage to the main road they can despatch goods to all parts of the country. The bricks are a fine bright red colour, very hard and dense and do not absorb so much moisture as those made from clay of a more soft, sandy nature. They are very durable and impervious to the weather and are used by leading architects, landowners, farmers and builders in Hereford and adjoining counties. The main part of the works is carried on in a large building 150 feet by 40 feet equipped with machinery including winding gear for drawing the wagons of marl and clay from the cuttings up an iron tramway to huge grinding rolls. There are five large kilns used for burning bricks and pipes and a smaller one for firing ornamental bricks, pottery and other high-class work.'[65]

The company list some of their more prominent customers: Barr's Court Railway Station and the goods station at Moorfields in Hereford, and parts of the Dinmore Tunnel for the railway companies; Credenhill Court Mansion for W. Farrer Eckroyd; alterations at Goodrich Castle; the building of Thinghill Court, Withington; Hereford County College and the Lunatic Asylum at Burghill; The Gas Works at Widemarsh, and the Water Tower at Broomy Hill for the Hereford Corporation. Godwin and Hewitt used the tiles and the bricks to build their Victoria Works, (next entry HSM 9414).

In 1858 two brickworks are listed at Holmer making 450,000 and 400,000 bricks respectively. The freeholders were Rev. J. Balmer and T. Tunks and the Manufacturer was Mr. Bristow. This we assume was one of them and Brickworks, Holmer, (HSM 31967) the other.

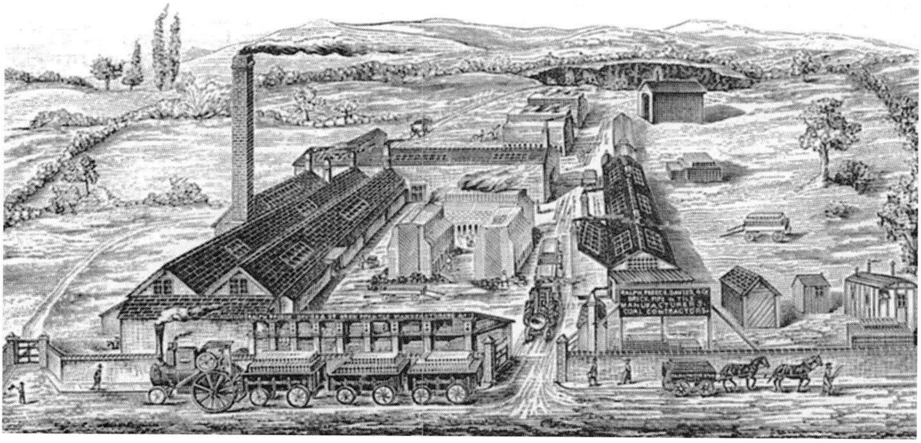

The Albert Brick and Tile Works in 1892

The company was also a haulier and traded in coal, coke, salt, lime and agricultural manure, with depots at the stations at Barr's Court, Leominster and Moreton-on-Lugg, as well as boasting 'special deliveries by traction power'. They delivered coal on 30 June 1904 to Bringsty and Hampton Court, probably for the Hampton Brickworks.[66]

'The 1886 map shows 12 brick factories here. ... I filmed the Holmer chimney stack being blown up in 1968-1969'.[67]

Site Name: **Victoria Tile Works**
HSM : 9414 GR: 5170 4180
Description: The Tile Works was started in 1878 by Henry Godwin after he split from his brother William, and sited on the premises of the former Herefordshire and South Wales Agricultural Manure Co. It lay between the railway and the canal off College Road. On the first edition OS map the brickyard is called the Victoria Tile Works. Later maps showed the works had expanded considerably to cover over 2ha.

Though this site was a tile works, not a brick works it is included since it has the best, and only, extant buildings in the county used in the manufacture of tile, which also illustrates the manufacture of brick, a similar process.

'Since WW2 ... [it has been] occupied by Herefordshire Whitestone Tiles Ltd, who have recently vacated it ... [and] many of the original impressive red brick buildings remain'.[68] Listed as Godwin and Hewitt Ltd., Victoria Tile Works, in the 1917 Kelly's *Directory* it was described as the largest in the city and county.

'In a hollow surrounded by charming scenery there is a long line of buildings including a chimney stack of 160 feet. Situated on six acres of clay beds the works employ a large number of men, women and boys in various departments. Famous for splendid examples of enamelled and embossed tiles with a wide variety of styles and colours. Themes are the British Flora and Fauna and from abroad. Especially famous for fire places which are sold throughout England, the Colonies and America.'[69]

By 1900 the Godwin businesses had either been sold or closed, but it was not until 1980 that the tile works were completely closed. By 2000 it was the site of the Holmer Industrial Estate. See also the box on the Godwin Family, page 48.

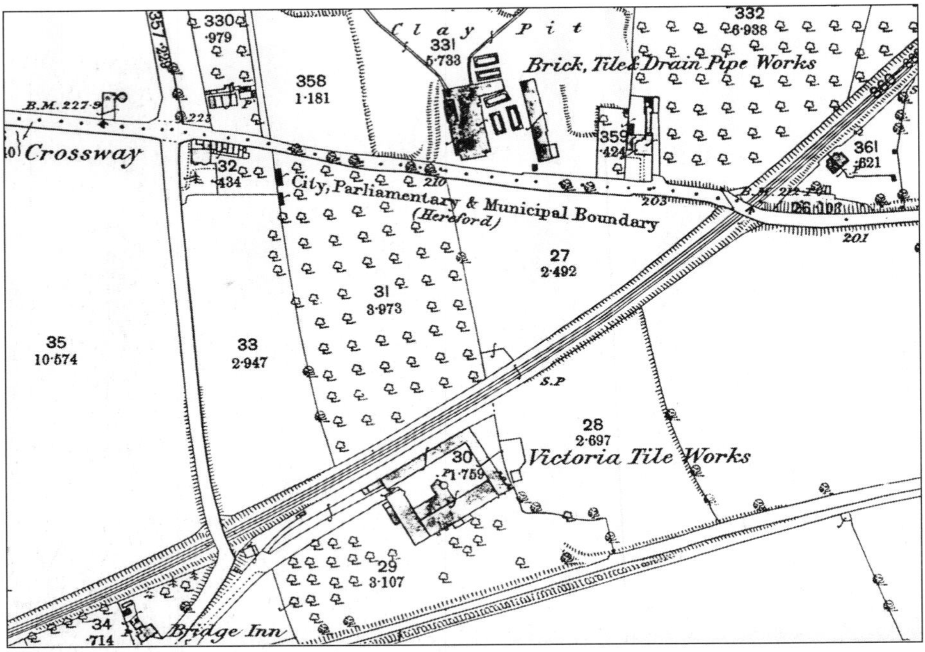

Top: OS map of 1888 showing the Albert Brickworks and the Victoria Tile Works.
Bottom: Aerial view of the Victoria works in its heyday

The Victoria Tile Works
Top: Present day view across the railway line
Bottom: Men at work

Site Name: **Worm Hill Brickworks**
HSM: 32794 GR: 5200 4200
Description: Brickworks whose precise location is unknown. In the mineral statistics of the Geological Survey it is listed as being owned by Tom Beech, the manufacturer being Thomas Tunks, and 500,000 bricks were made per annum from red marl.

The only Wormhill in the vicinity is across the river at Eaton Bishop. While this is low lying and leads to what was possibly a ferry crossing in the 19th century along the Roman road it would seem strange for the Geological Survey to list it as being in Holmer.

Site Name: **Brickworks, Holmer**
HSM: 31967 GR: 5200 4200
Description: In 1858 two brickworks were present in Holmer, the other we assume was the Albert Steam Pipe and Brickyard HSM 32329. One freeholder was Rev. J. Balmer the other T. Tunks. The manufacturer of both was Mr. Bristow. One made 450,000 brick per year, one 400,000. As near as we can estimate this one was situated between the A49, A4103 and Attwood Lane and now built over.

Hope under Dinmore

Site Name: **Brickworks near railway tunnel**
HSM: 34275 GR: 5040 5215
Description: A brickworks is shown on the 1889 map, with one small building. On the tithe map it is not present and the field is called Meadow. On the 1964 map only a track is present. A 100m spread of brick and charcoal located at the point where the field levels off towards the brook is probably associated with two brickworks.

The first brickworks appears to have been built by the 1852 Shrewsbury and Hereford Railway Company and was associated with the 1,000-yard railway tunnel, opened in 1853. This brickyard was abandoned between 1889 and 1905 and a new one built about 120m north of the old site, immediately south of Cherry Brook. (Cherry Brick Orchard, HSM 32445, is about 1km away to the north-west). A roadway (50536 52261 NNW-SSE) with wheel ruts may have been another access route to the brickworks. A further feature at 50554 52293 was interpreted as a short-lived tram road to the new brickworks.[70] Sample bricks can be found at the site.

Site Name: **Hampton Court Estate 19th century Brickworks**
HSM: 32845 GR: 5210 5240
Description: Evidence from Hereford Record Office (E41/326 OL365) gives the office and accounts for Hampton Court Brickyards. These show what was made and that the products were sold mainly to firms and farms. The quantities were normally in the tens and hundreds of thousands. Bricks, pressed bricks, drain pipes (agricultural) of 2 inch to 12 inch diameter, quarry tiles, tiles, coping and gutter bricks were all made. They were operating between 1871 and 1880. See also Cold Oak Brickworks, Newton (Leominster) HSM 32283.

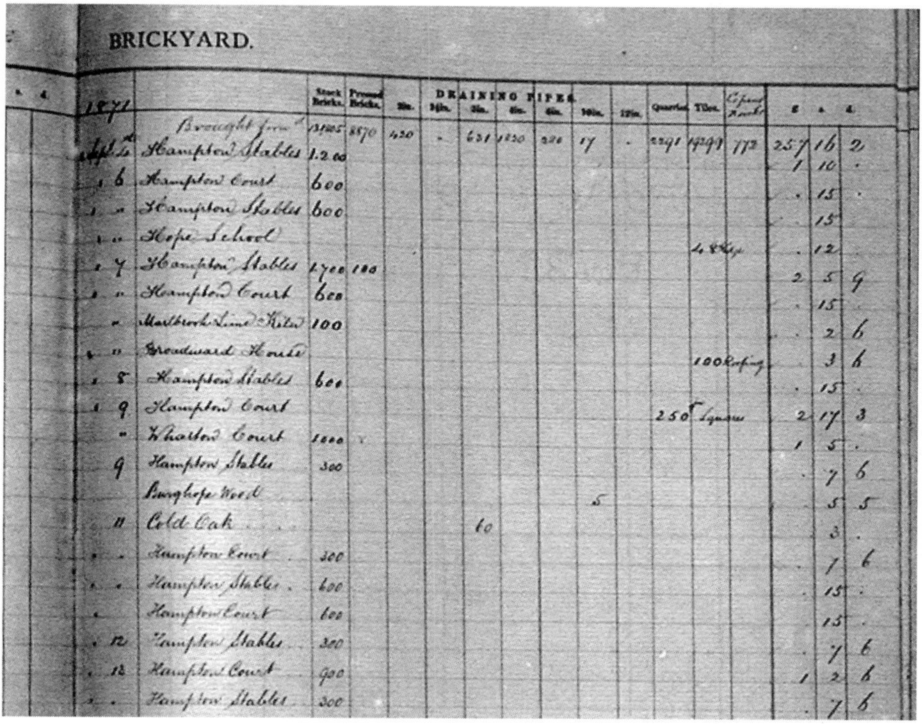

BRICKYARD.

1871		Stock Bricks	Pressed Bricks	Other	£	s	d
	Brought forward	[illegible]	8870	420; 630; 1230; 220; 17; 2291; [illegible]; 772	257	16	2
Sept. [illegible]	Hampton Stables	1200			1	10	.
6	Hampton Court	600			.	15	.
"	Hampton Stables	600				15	
"	Hope School			48 [illegible]	.	12	
7	Hampton Stables	1700	100		2	5	9
"	Hampton Court	600			.	15	
"	Marlbrook Lime Kiln	100			.	2	6
"	Broadward House			100 Roofing	.	3	6
8	Hampton Stables	600			.	15	.
9	Hampton Court			250 [illegible] Squares	2	17	3
"	Wharton Court	1000			1	5	.
9	Hampton Stables	300			.	7	6
	Burghope Wood			5		5	5
11	Cold Oak			10		3	.
"	Hampton Court	300			.	7	6
"	Hampton Stables	600			.	15	.
"	Hampton Court	600				15	.
12	Hampton Stables	300			.	7	6
13	Hampton Court	900			1	2	6
"	Hampton Stables	300			.	7	6

A page from the Hampton Court Brickworks Account Book

Site Name: **Cherry Brick/Brook Orchard**
HSM: 32445 GR: 5130 5305
Description: This is now an arable field lying along the A417 between the A49 and the River Lugg.

Site Name: **Hampton Court Estate 20th century Brickworks**
HSM: 32849 GR: 5200 5200
Description: Documents from the Cale family of Burley Gate given to the Bromyard History Society have records of Brickworks at Hampton Court dating to 1903. These show that thousands of bricks were produced. On 30 May for example 13,000 bricks at a cost of £3 13s 6d were inventoried. The total between 30 May and 13 August was 48,000 bricks.

Further 1903 records give daywork dues for digging clay in March as £1 3s 0d and for preparing floors on 23 May as 11s 0d. A summary of the account shows in 1903 the cost of brick making was £13 7s 1d plus day work at £7 13s 10d and coal at £23 5s 9d. The value was £67 4s 0d, once expenses and preparing dry ground and hauling were deducted it left a profit of £21 10s 10d.[71] See also Cold Oak Brickworks, Newton (Leominster) HSM 32283.

Site Name: **Clay Pitts**
HSM: 25471 GR: 4840 5302
Description: This is a long way from any of the brickworks in this parish and we were unable to positively identify a nearby brick activity.

Humber
Site Name: **Brickyard**
HSM: 32446 GR: 5440 5517
Description: Sited next to Risbury Court.

Kentchurch
Site Name: **3 Clay Pits**
HSM: 41352, 41351, 41354 GR: 4297 2857, 4296 2864, 4290 2840
Description: We have been unable to identify any brick activity in this parish. However, these three clay pits, taken from the first edition of the OS map, may be all that remains of single use sites.

Kimbolton
Site Name: **Brick Kiln Piece**
HSM: 32331 GR: 5050 6061
Description: Sited beside the canal at its southern terminus.

Site Name: **Brick Piece**
HSM: 36531 GR: 5150 5990
Description: There is little evidence of crop marks, though there is some disturbance in the south-east corner of the field which has the tithe number 555.

Site Name: **Brick Piece**
HSM: 32330 GR: 5170 5989
Description: This one has the tithe field number 852. It is a large field reaching from the minor road to the brook.

Kings Caple
Site Name: **Brick Moor Rise**
HSM: 32334 GR: 5740 2850
Description: Sited beside River Wye to take advantage of the easy transport.

Site Name: **Brick Kiln Rises**
HSM: 32333 GR: 5515 2990
Description: Sited beside River Wye to take advantage of the easy transport.

Site Name: **Brick Kiln Orchard**
HSM: 32332 GR: 5595 2979
Description: Situated well above the river and beside minor road and able to serve the village of Kings Caple.

Kings Pyon

Site Name: **Brick Kiln**
HSM: 18752 GR: 4990 4290
Description: Located south-west of Wistaston.

Kingsland

Site Name: **Brick Close Orchard**
HSM: 32447 GR: 4457 5915
Description: Close to three farms and able to serve any or all.

Site Name: **2 Clay Pit**
HSM: 41050, 41051 GR: 4257 6087, 4270 6021
Description: Identified from the first edition of the OS map, we have been unable to positively identify suitable brick activity in Kingsland.

Kingstone

Site Name: **Brick Clamp, Brick Clamp Piece**
HSM: 32336 GR: 4413 3519
Description: This site was probably located on the common at the time. Note that 300m to the north there is an enclave of this parish in the next one, Allensmore, of two small fields 480, 481, today with ponds; these may have a source of the clay. The fields are un-named in the tithe survey.

Kington

Site Name: **Brick Kiln Common**
HSM: 32335 GR: 2876 5515
Description: Located on top of a hill on common land and 200m north-west of Bredward Brickworks, HSM 31581, which is dated to pre-1890. Possibly there were long term brickmaking businesses on this site. Fuel would have come from the steep valley sides, which are still partly wooded.

Site Name: **Brick Works**
HSM: 31589 GR: 2895 5495
Description: These Brickworks are not shown on the tithe map, but the cottages to the south of it are. They are not specifically named as Brick Kiln Cottages, though they are on the 1964 map. On the first edition of the OS map the brickworks are shown with a clay pit and the cottages, which are again not named.

Site Name: **Flintsham Brickworks**
HSM: 32487 GR: 3250 5895
Description: Flintsham was a comparatively large brickworks in 1858. It was one of 25 mentioned by the Geological Memoir. Bricks were made from the clay of the Old Red Sandstone. The land owner was the Right Hon. Jane Lady Langdale. The manufacturer was John Meck Milner. 330,000 bricks were made per annum and 300,000 draining pipes. On the tithe map a brickyard is shown to the east of Flintsham Square on the south side of the road.

Site Name: **Bredward Brickworks**

HSM: 31581 GR: 2892 5500

Description: Brickworks at the top of a hill above Red Wood with associated clay pits recorded at 28970 55030. The area is now partially wooded and the field boundaries have been lost (as shown on OS Explorer 201). Marked as Brickworks on the first edition of the OS map, information concerning them is very limited. Brick-kiln Common is marked on the OS map of 1833. The 1881 census names Llewellyn Hancock as a brick- and tilemaker of Kingswood. In 1900 the brickworks were not working, though many of the cottages in Kingswood were built from the local bricks. Until quite recently the old brick kilns could still be seen on the site.[72] Marshy areas are the only indication of the works.

Site Name: **Clay Pit**

HSM: 41074 GR: 2897 5503

Description: Although we have been unable to positively identify brick activity, the site is close to Brick Kiln Common, HSM 32335, and Bredward Brickworks, HSM 31581.

KINNERSLEY

Site Name: **Kinnersley Brickworks**

HSM: 31566 GR: 3357 5026

Description: This brickworks and clay pit with a kiln is marked on the 1891 OS map. Situated beside the Kinnersley/Almeley road. A small pool remains on the modern OS Explorer 201.

Site Name: **Clay Pit**

HSM: 41168 GR: 3430 4885

Description: This clay pit may have been used by Kinnersley Brickworks, HSM 31566, or have been dug for a single use. It was identified from the first edition OS map.

KNILL AND LOWER HARPTON

Site Name: **Brick Grounds**

HSM: 32337 GR: 2888 6025

Description: Now located on meadow land.

LAYSTERS

Site Name: **Laysters Brick and Tile Works**

HSM: 31583 GR: 5605 6357

Description: The brickworks at Laysters Pole are 200m east-south-east of the A4112 by the junction of Bockleton Road and Great Heath Wood. In 1842 the land was owned by Thomas Hutton and occupied by William Handley. The three relevant fields are Cottage Meadow, which is the site of the brickworks and its clay diggings, Cottage and Garden, and Feggry Piece, into which another clay pit has been dug. These formed a triangle 150m by 300m (2.25ha) bounded on the east and west sides by roads, and the south, between the site and the claypit, by a track. The triangular area is generally 0.75m lower than the surrounding land where the clay has been extracted, descending to 1.5m in the centre. It is wet and marshy with secondary woodland of about 10-25 years growth, some of it coppiced.

By 1861, when the population of the village was 283, the brickworks were established. The owner was William Lort of Kings Norton, and his steward was John Callow, who was also an agent for the Norfolk Farmers Livestock Insurance Co, and for the Lancashire Fire and Life Insurance Office. Ten years later the population of the village had fallen marginally to 279, and the landowner was the Trustees for Mrs. Catherine Lort and William Lort of Yockleton Hall near Shrewsbury. Mrs. Lort had by now become the proprietor of the Laysters Brick and Tile Works, and John Callow continued as her agent, and also for the two insurance companies. In the 1881 census, Joseph Band, living at the Brickyard, and William Nash, living at Pole Lane, were both described as brickmakers.

Charles Cooke of Middleton-on-the-Hill was a labourer at a brickyard, probably this one. The population that year had fallen further to 238, and in 1885, Mrs. Lort had become a private resident at Great Heath, Laysters, and she was described also as a farmer and hopgrower at Great Heath Farm, where the present landowner, Mrs. Cathie Wilson lives.[73]

Today, at the south end just to the south east of the ruined cottage (field number 186), there is a circular mound 9m in diameter, 2m in height, with a flattened top slightly dipping in the centre, with many bricks and pipes on the surface. One line of bricks is probably structural. Badgers have been digging out on its west side. This is interpreted as a kiln of bee-hive type. The cottage shown on the tithe map is today a ruin. The one standing wall is of stone rubble with a gable end about 3.5m in length. Attached to this is a brick cellar about 2m by 3m. The cottage we assume pre-dates the brickworks, and was repaired and extended in brick.

Various linear banks and small ponds lie in the vicinity, suggesting clay digging was carried out over a larger area. To the east, 100m across a field there is another pond, but this is deeper. It is shown on the tithe map and may not be associated with the brickworks.

The remains of the kiln at the Laysters Pole Brickworks, now very overgrown. Indeed, several old sites across the county are now hidden by such scrub

Site Name: **Clay Pit**
HSM: 41066 GR: 5597 6348
Description: Since this pit was identified from the first edition OS map, it may have been used by Laysters Pole Brickworks HSM 31583, as it lies 600m to the south-west.

Lea
Site Name: **Brick Kiln Field**
HSM: 32450 GR: 6525 2250
Description: Located beside the B4224.

Ledbury
Site Name: **Fairtree Farm Brick and Tile Works**
HSM: 35553 GR: 6935 3750
Description: The tithe map of 1841 has field 1669 as Brickyard and adjoining field 1212 Great Meadow. On the 1885 map it has shifted to the south a little at 6950 3730 and is an operational brick and tile works (the 1840 location is shown as an earthwork). By 1964 there is nothing left at the site except a depression in the ground, though it covered a few hectares. Today there is a fisherman's lake. The soil is a red terracotta colour. It is likely that it subsequently served the Ballard Brick Works (HSM 32792) as a clay pit.

Site Name: **The Elms Brickyard**
HSM: 35554 GR: 7180 3815
Description: Field number 148 is called Brickyard on the tithe map and is called Old Quarries of the first edition OS map. This site visit revealed evidence of earth digging in that the field adjacent had been lowered by 0.5m to 1m. The bricks in a standing wall nearby were 205mm to 220mm by 67mm to 74mm by 105mm to 110mm, with rounded coping stones.

Site Name: **Brick Kiln**
HSM: 20101 GR: 7134 3811
Description: Located at Holborn Croft. A small overgrown area at the top of the field adjacent to Green Lane to the east has a hummock, which is possibly the site of a brick kiln referred to in 1672 in the Diary of George Skippe: 'I paid ... for brick for finishing my brick kill [*sic*] in Holborn Croft'.

Site Name: **Brickyard**
HSM: 19856 GR: 7065 3735
Description: In Victoria Road, and marked on the 1841 tithe map. Perhaps the same site as the Brick Kiln marked on the 1885 map just to the west of Victoria Road. This may have been the brickworks listed in the 1858 Mineral Statistics of the Geological Survey which was called Quarry Yard. Bricks were made from yellow clay, the land was owned by Earl Somers and the manufacturer was Mr. Reece. Output, of bricks and pipes, was about 400,000p.a. It was a sports field in 2003.

Site Name: **Ballard Brickworks**
HSM: 32792 GR: 7025 3675
Description: The present site of Ledbury Preserves was Ballard Brickworks before the First World War and was owned by Robert Ballard. J.G. Davies was related to the owners, and the Fairtree Farm opposite was run by the Davies family.[74] Clay was dug on the spot and probably Fairtree Farm. (HSM 35553).

The Ballard family in Ledbury were famous both locally and abroad. Thomas Ballard was the eldest son and he painted the Last Supper which is behind the altar in Ledbury church. Stephen Ballard was involved in the building of the Hereford to Gloucester canal. The town of Ballard in north Queensland, Australia, is named after another son, who was a railway engineer. Ballard bricks made the Ledbury Rail Viaduct (see pp.62-63 for more information about the Ballard family), and some were sold to the Canal Company. Stephen Ballard made a machine for cleaning bricks in 1840-1850, enabling used bricks to be used again. Stephen Ballard was an ancestor of Robert Barns who has archives of the Ballard family going back to 1590.[75]

'My grandfather died in 1931, Ledbury Preserves was a brickworks and my family lived in Fir [*sic*] Tree Farm opposite. Hill House was made with Davies bricks, stamped with J C D. Pixley Rectory and houses on Bye Street also used Ledbury bricks. When grandfather was 15 clay came to Lebury on tractor engines brought down from the railway, so some clay was imported. Once the steam engines frightened horses pulling a coffin which ended upside down on the pavement.

'In 1858 R. Ballard made 500,000 common bricks, tiles and pipes *per annum*. The source is called marly clay and the land was owned by R. Biddulph.'[76]

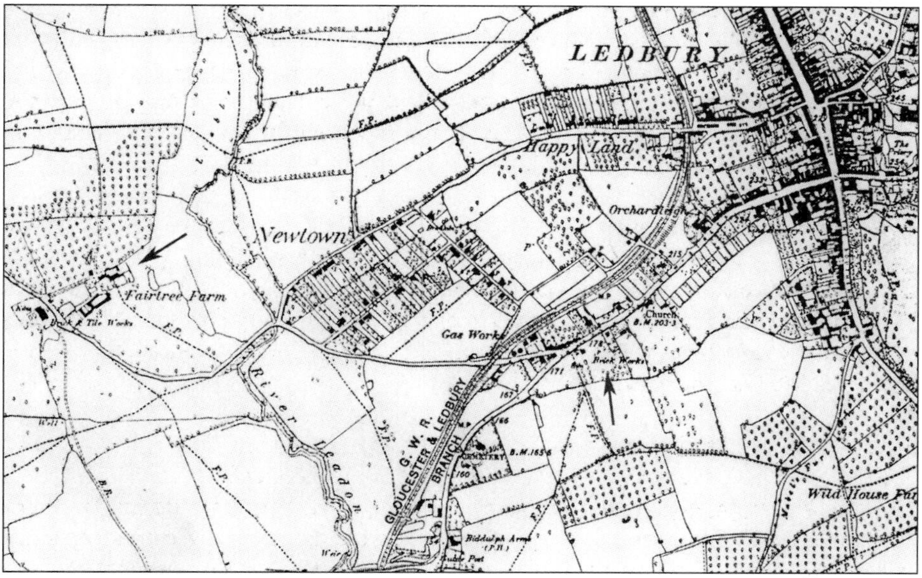

In 1886 both the Ballard Brickworks and the tile works at Fairtree Farm were in evidence on the OS map (Hereford 41NE)

Site Name: **Brickworks**
HSM: 33851 GR: 7170 3828
Description: The 1917 Kelly's *Directory* lists a brick and tile makers as George Hill and Sons, at the Homend Ledbury. Nothing shows here on the maps, though on the 1964 OS map there is a kiln in a quarry, off the Worcester Road. This is possibly a lime kiln however at Worcester Walk Coppice. On the opposite side of the road to this kiln is The Elms Brickyard (HSM 35554).

Site Name: **Brickworks**
HSM: 19845 GR: 7065 3723
Description: A brickworks is shown on 1887 map as still working with a kiln and other small buildings adjacent on the south side to New Street. The site was visited in 2003, and the last houses in Oatley Road are set 1.2m below surrounding land (see photograph on p.28).

LEINTHALL STARKES
Site Name: **Brick Kiln Field**
HSM: 32338 GR: 4467 7028
Description: By a minor road and about 1.5 km from the village.

LEINTWARDINE
Site Name: **Brick Close (Nearer, Farther)**
HSM: 32339 GR: 4057 7310
Description: On the Roman road south of the village, where there are extant ponds.

LEOMINSTER
Site Name: **Barons Cross Brickyard**
HSM: 32340 GR: 4811 5860
Description: This brickworks was located on the outskirts of town by the present A44, and is now built over and orcharded. It was one of 25 in the county mentioned in a Geological Survey of 1858. The material used was red marl, the owner was F. Davis Esq., and the manufacturer was Ann Tunks. Red bricks, roofing and paving tiles and drain pipes were made, as many as 300,000 per annum.

Site Name: **Brick Kiln Croft**
HSM: 18229 GR: 4800 5515
Description: Although it is in Leominster Parish it is nearly 5km to the south-south-east, and lies in a field 450m south of Ivington Park.

Site Name: **Brickyard Garden**
HSM: 32341 GR: 4935 5885
Description: This brickworks was located on either side of Dishley Street on the west side was Cattle Market, now a car park, and the east side is now built over. It was on alluvial clay and the manufacturer was Mr. Rogers. Only one brickworks is listed in the geological survey for Leominster, and although it may be this one, also see Barons Cross Brickyard HSM 32340.

Site Name: **Ebnall Brickworks**
HSM: 38763 GR: 4751 5895
Description: Ebnall Brickworks shown on the first edition OS map just east of clay pits. No buildings are marked. This may have been the Ebnal (*sic*) New Brick Co., listed in the 1917 Kelly's *Directory*.

Site Name: **Site of Brickyard**
HSM: 19526 GR: 4940 5890
Description: At Dishley Street and shown on 1835 map, this brickyard had gone by the 1887 first edition OS map and had been replaced by a hop yard. The 1835 map shows features which may be a kiln to south of site. In Kelly's 1891 *Directory* Llewellyn Hancock of Bargates Leominster is listed as a brick- and tilemaker. This may also be the brickworks listed in the 1858 mineral statistics. It was owned by John Rogers and the manufacturer was also John Rogers and described as being at Stocking Hill. John Rogers owned land at Upper Hide on the 1840 tithe map, but a brickworks is not mentioned.

Site Name: **4 Clay Pits**
HSM: 41093, 41090, 41089, 41092
GR: 4755 5939, 4905 5828, 4843 5883, 4751 5900
Description: We have been unable to positively identify brick activity in relation to these sites. However, 41093 and 41092 may have been one of the pits referred to for the Ebnall Brickworks HSM 38763. Sourced from the first edition OS map.

Letton

Site Name: **Waterloo Brickworks**
HSM: 34072 GR: 3410 4755
Description: Brickworks (disused) are shown east of the road at Waterloo on the 1885 first edition OS map. Two clay pits lie adjacent (see HSM 41169 and 41170), these are ponds on the 2000 map. This area was a common on the Letton 1840 tithe map. On the 2000 map there are several additional lakes/ponds not present on the 1885 map. The pond is permanent and groundwater fed.

Site Name: **2 Clay Pits**
HSM: 41170, 41169 GR: 3420 4756, 3401 4757
Description: These are the claypits mentioned in Waterloo Brickworks HSM 34072.

Lingen

Site Name: **Brick Kiln**
HSM: 21906 GR: 3625 6951
Description: The nearby farm at Birtley is called Brick Kiln Farm. Wasters can be seen in side of the road at SO 3615 6948 and there is a small pond that probably derived from the brickmaking. The owners found bricks and stones, and kiln waste was encountered in the field called the Orchard on the tithe map (No 18). Notes by George Marshall tell of a man called Tailor who, if still alive would have been 90 in

1945. When he was 8 years old he trod the clay with bare feet. Mr. Lawdry was the brickmaker, he dug the clay while his wife moulded the bricks. This would have been in 1855. Probably the Methodist chapel was built of these bricks.

Site Name: **Brick Kiln**
HSM: 36580 GR: 3626 6760
Description: Local people told me that bricks were made here and brick wasters can be seen in the fields by Marsh Cottage, Birtley.

Linton (Bromyard)

Site Name: **Brick Clamp Piece**
HSM: 32342 GR: 6614 5454
Description: Located to the south of the old railway, it is now under the sewage works. Note Brick Barns on north side of railway.

Site Name: **Clay Pit**
HSM: 41155 GR: 6939 5371
Description: We have been unable to positively identify a brick activity nearby. Sourced from the first edition OS map.

Linton (Ross on Wye)

Site Name: **Brick Kiln Meadow**
HSM: 32343 GR: 6695 2626
Description: Now under the M50.

Little Hereford

Site Name: **Upper Upton Farm**
HSM: 32846 GR: 5518 6614
Description: Hand made bricks were found on this site. Several had stuck together in a clamp from what would have been a small brickworks, and probably used for the farm's own requirements. Tile found on site measure 290mm by 160mm by 17mm, red in colour with a rough texture. The brick was 235mm by 115mm by 80mm with a smooth texture and terracotta colour. One brick has a pawprint, one some white mortar.

Site Name: **Clay Pit**
HSM: 41037 GR: 5405 6627
Description: We have been unable to positively identify a brick activity nearby. Sourced from the first edition OS map.

Little Marcle

Site Name: **Brick Kiln Orchard**
HSM: 32344 GR: 6620 3605
Description: The only evidence, apart from the name is a pool in middle of the field sited 300m south-west of The Court.

Llanwarne & Llandinabo

Site Name: **Brick Fields (Llanwarne)**
HSM: 32345 GR: 4915 2805
Description: Located between Hill Farm and Hill Gate, and probably served both. Note the pond in next field to the east which may be the original claypit.

Site Name: **Town Pitts, Upper Clay Lands, Wood Pile Orchard**
HSM: 32346 GR: 5143 2882
Description: This is a possible site for brickworks. Field no. 46 (Town Pitts) is next to 47 (Upper Clay Lands), and 77 Wood Pile Orchard is across road from 47.

Lower & Upper Bullingham

Site Name: **Brick Kiln**
HSM: 25445 GR: 5170 3580
Description: Identified from the field name 'Brick Kiln'. The next field to the north is called Wood Meadow suggesting a source of fuel. The site in 1887 on the first edition OS map is marked as rough ground with a track leading to the road from the north-east corner of the site. In 2000 the area was wooded.

Site Name: **Brick Clamp**
HSM: 32348 GR: 4960 3550
Description: This site is 700m south of the Grafton Inn (formerly Brick Kiln Inn SMR 31557, Grafton).

Site Name: **Brick Meadow**
HSM: 32347 GR: 5090 3661
Description: Located 500m south of Bullinghope village.

Site Name: **Brick Meadow**
HSM: 32438 GR: 5295 3725
Description: As near as we can estimate the site is most likely built over by Woodlands Farm.

Lugwardine

Site Name: **Godwin Tile Works (1st), from 1849**
HSM: 11862 GR: 5510 4130
Description: Brickworks & tile factory was established by William Godwin 1848 at Hemhill. The offices still stand and are now converted into two cottages. Behind them is a row of low brick built sheds which were originally workers houses. Brickyard Cottage was rebuilt in 1990s from original bricks on site. Fairview House is the original site building, by 2003 it was a dwelling.

A small brickworks had operated in Lugwardine since the early 1800s, William Godwin, a brickmaker previously working in Ledbury, bought the lease in 1849. He employed ten labourers. Henry Godwin joined his brother in 1852 after working in the Worcester Porcelain Company and the company began to make decorative tiles.

The business was very successful and William Godwin was able to buy Porch House, Lugwardine with a considerable amount of land, including also the freehold of the tile works. In 1861 he bought the Withington works (SMR 11863 in Withington parish).[78] For a description of the business see the box on the Godwins pp.48-50.

Luston

Site Name: **Brickton Croft**
HSM: 32349 GR: 4765 6395
Description: A possible brickworks but we have no information about it.

Lyonshall

Site Name: **House and Garden**
HSM: 32451 GR: 3155 5520
Description: The house was owned by a brickmaker and when clay gave out he moved on.[79] Subsequently the house became dilapidated and the stone was taken about 25 years ago. On the OS first edition map.

Site Name: **Brick Kiln Piece**
HSM: 32452 GR: 3430 5840
Description: Now partly covered by the dismantled railway with a pond in the field to south of railway.

Site Name: **Upper/Lower Bricklands**
HSM: 32453 GR: 3240 5510
Description: Only represented by a pond in the north-west corner of a field.

Madley

Site Name: **Bricknall Close, Meadow**
HSM: 32351 GR: 4070 3895
Description: Situated on the north side of B4352 and 600m west of Showell Farm. Bucknalls Wood is the next plot north, and could have been used for the fuel. Bricknall and Bucknalls are sufficiently close in spelling to suggest a possible copying error.

Site Name: **Brick Close**
HSM: 32352 GR: 4035 4127
Description: Located close to Lulham House and Lulham Court Farm, it could have served either.

Site Name: **Brick Clamp Piece**
HSM: 32350 GR: 4055 3625
Description: The field is 300m south-west of Great Brampton. All that is left is a pool in the field.

Site Name: **Brickyard**
HSM: 32440 GR: 4217 3790
Description: Located to the north-west side of the Roman Road.

Site Name: **Field and Brickyard adjoining the Brickyard**
HSM: 32353 GR: 3907 3822
Description: This site has a more positive name than so many others, and it is 500m west of Lower Shenmore. Note strip fields to the east, numbered 1014, 1015, 1016 which are water today.

Site Name: **Brick Meadow**
HSM: 32439 GR: 4236 3660
Description: This site is 1km from Kingstone village centre, close to Webton Court.

Mansell Lacy

Site Name: **Westmoor Brickworks**
HSM: 31570 GR: 4108 4552
Description: This is a large site comprising two buildings, 100m apart, lying north-west/south-east of each other. Today a small pool is shown in the north-east corner of the site, and there is damp ground on the west side. A visit in 2003 revealed undulating ground planted with crops.

Site Name: **Brick Kiln Field**
HSM: 32354 GR: 4279 4281
Description: This site lies next to the Roman road from *Magnis* to Bishopstone.

Marden

Site Name: **Brick Close**
HSM: 32356 GR: 5468 4778
Description: Lies next to Amberley Court, and was probably for the Court's own use.

Site Name: **Brick Kiln Piece**
HSM: 32448 GR: 5310 4650
Description: Near a bridleway on the southern boundary of the field.

Site Name: **Brick Lump Orchard, Brick Orchard**
HSM: 32355 GR: 5318 4987
Description: On the north slope overlooking the Lugg Valley which here is a glacial water spillway. There are two small ponds on the southern boundary of the field shown on the OS Explorer map 202. (Brick Lump = Brick Clamp?).

Mathon

Site Name: **Bell Inn Brickworks**
HSM: 35555 GR: 7540 4470
Description: A brickworks is marked on the 1885 first edition OS map just to the south of Bell Inn. It is shown as a kiln, buildings and quarries, though brickworks are not marked on the tithe map here. In one field to the north-east an 'old clay pit' is shown. On the 1964 map a few small buildings and earthworks are still present, but not the brickworks. A site visit in 2003 revealed that the ground has been filled in the last 20 years with builders' rubble and is now used as grazing for horses.

Site Name: **Brick Kiln Piece**
HSM: 32455 GR: 7582 4635
Description: Close to West Malvern about 500m away and below a wooded hill.

Site Name: **Brick Kiln Piece**
HSM: 32358 GR: 7340 4580
Description: Located between Colwall and West Malvern.

Site Name: **Brick Kiln Meadow**
HSM: 32357 GR: 7418 4365
Description: There is a pond on the southern boundary and a building on the north-eastern boundary, both shown on the OS Explorer map 14.

Site Name: **3 Clay Pits**
HSM: 41301, 41299, 41298 GR: 7359 4430, 7625 4470, 7557 4484
Description: We have been unable to positively identify a brick activity nearby, and 41301 and 41299 are not the clay pit referred to in Bell Inn Brickworks HSM 35555, though 41298 probably is.

Michaelchurch Escley

Site Name: **3 Clay Pits**
HSM: 41310 41309 41311 GR: 3153 3750, 3034 3748, 3072 3583
Description: We have been unable to positively identify a brick activity near any of these clay pits. Sourced from the first edition OS map.

Middleton on the Hill

Site Name: **Earthworks, Redwood**
HSM: 30261 GR: 5650 6532
Description: A semi-circular feature, the site consists of an earthwork on the east and cropmark on the west and is apparently a hollow. A clear track leads to Redwood farm. On the OS Explorer 203 a stream is shown running down a steep hill south to north. It has been channelled into a straight ditch adjacent to the site and continues as a natural feature down the hill. North of the site field 134 is marked as a clay pit. The site itself is shown as a small oval feature, probably a pond with a straight ditch adjacent and has been interpreted as a clay pit with/or an industrial site.

Site Name: **Brick Kiln**
HSM: 19111 GR: 5435 6520
Description: The field name is Brick Kiln Piece and Brick Kiln, north-east of Nurton Court, Middleton.

Moccas

Site Name: **Brick Kiln**
HSM: 32359 GR: 3528 4301
Description: Adjacent fields named for tithe field no 21 Brick Kiln, 20 Part of Brick Kiln, 36 Brick Kiln Grove; also adjacent possibly for fuel supply: 41 Coppice, 32 Wood adjacent to Croshand (now Cross End) Coppice. These are also shown on 1772 estate map as brick kilns. Moccas Court is built of brick and it maybe that this kiln supplied them. See also p.62.

Mordiford

Site Name: **Clay Pit**
HSM: 41373 GR: 5869 3724
Description: We have been unable to positively identify a brick activity nearby. Sourced from the first edition OS map.

Much Birch

Site Name: **Brick Close Meadow**
HSM: 32442 GR: 5040 3035
Description: Lies 100m south of Court Farm and 200m from the village.

Site Name: **The Brick Piece**
HSM: 32441 GR: 5135 2995
Description: Situated 200m from the A49 and adjacent to a footpath and minor road.

Site Name: **Clay Pit**
HSM: 41378 GR: 5155 3001
Description: We have been unable to positively identify a brick activity nearby, though see The Brick Piece HSM 32441 which may be linked. Sourced from the first edition OS map.

Much Cowarne

Site Name: **Burley Gate Brickworks**
HSM: 31569 GR: 6000 4664
Description: Brick and tile works on the south side of the present A417 and 100m west of Orls Farm. The south side of the site is now wooded (OS Explorer 202), and this is where the claypits were sited.[80] There is a brick and tile scatter. The bricks are large and have frogs. On the tithe map an extensive area of clay pits are marked. A wonderful photograph shows men and boys moulding bricks, stacking them out to dry and covering them with straw and this photograph is shown on p.32. The field is open, level, probably dug down one metre all over, and hedged. The Bromyard History Society holds work books for the Burley Gate Brickworks.

Barry Lawrence writes: 'I located a very dilapidated volume of 286 pages which is the Sales Ledger from the Old Brickyard at Burley Gate and held by the Bromyard History Society. It covers the years 1881 to 1899, and it lists the items sold and the names and addresses of the recipients of the brick products.

'The brickyard was started in the 1870s by William Cale and continued to within living memory; so by the time of the ledger the firm was well established. Among Mr. Cale's main customers were Henry Higgins of Thinghill Mansion who died in 1890 and later his wife Elizabeth who died in 1898, and also Col. Bourne of Cowarne Court. Between them these individuals owned the majority of the properties in the district and were responsible for their upkeep or indeed, their original erection. Mr. Cale obviously had his own bricklayers as many of the entries relate to bricks 'sold' to 'own team'. Unfortunately it does not say where they were used.

'Amongst the more interesting entries are the ones relating to the school at Burley Gate (now the Village Hall) for which Mr. Cale provided 29,600 bricks between June and December 1889 at a cost of £44 8s (equivalent to 0.36d per brick or in today's money 6.6 bricks per 1p) and paid for by Mr. Higgins. The School House took 24,500 bricks during 1890 and 1891, whilst in 1896 the Post Office probably had trouble with the drains as fifty 2 inch and fifty 3 inch pipes were delivered at a cost of 3s 6d.

'At the same time Col. Bourne was involved in expense at 'The Elms' which used up 30,600 bricks at 0.33d each – were they inferior to those used at the school or was Col. Bourne more favoured as he was also Mr. Cale's landlord? Previously, from 1892–1894, William Amos, the new tenant at 'The Farm', must have been draining his land, as he used 11,300 pipes of different sizes ranging from 0.25p to 0.84p each. He also used 26,700 bricks, so perhaps there was a rebuild or a major extension. On a more modest scale some of the tenants obviously paid for their own material as there are entries, among others, relating to Mrs. Brace who lived at Crossways Cottage (200 bricks costing 6s) and also Mr. Hodges of the blacksmith's shop who in 1882 received 100 bricks and assorted pipes for 2s 6d.

'In addition to the immediate locality, Mr. Cale had among his more illustrious customers, Lady Emily Foley of Stoke Edith Park and the Lord Bishop of Hereford. He also supplied bricks and pipes to local builders including Mr. Charles Bufton of Ocle Pychard, and Messrs. Wood and Brothers who did much work in the Ullingswick and Felton area. It is a shame there appears to be just one volume. I wonder what happened to the others. Does anyone know?' A copy of a single page from this ledger is held at the HSM, and all the records can be seen at the Bromyard History Society Offices.

On 21 September 1940, under the Domestic Shelter Scheme, A.J. Francis, the Supervising Officer of the Surveyor and Waterworks Engineers' Department at Malvern Urban District Council writes to K. Kale [*sic*] Esq., of the Brickworks, Birley [*sic*] Gate, West Hereford: 'I thank you for your letter of 19th instant regarding the purchase of the Brick Kiln. It would be rather

The frog in one of the Burley Gate Brickworks bricks showing the gate trade mark

The Burley Gate Brickworks clay workings as they are today

difficult for me to come over at the moment, but I should be glad to purchase the kiln at the price mentioned at our recent meeting, ie £30.0.0 without the hut. I shall appreciate a reply from you regarding this matter at your earliest convenience.' It was presumably sold at this price.

Site Name: **Brickworks, Clay Pits and Brickyard Cottages**
HSM: 32793 GR: 6050 4750
Description: Shown on the 19th-century maps, this may have been the original site of the Burley Gate Brickworks (SMR 31569). Within this area on the modern map are Brickyard Cottages and Claypits Farm, and both are close to the Moreton Jefferies parish boundary. An 1858 geological survey lists a large brickworks at Moreton Jefferies, one of 25 in the county. The owner of the land was J. Morley Esq, the manufacturer was Timothy Edwards. The works make per annum 200,000 bricks, 50,000 drain pipes, 10,000 tiles, 10,000 pantiles and 1,000 crest tiles. This may be the location for this brickworks.

Site Name: **Brick Close**
HSM: 32360 GR: 6370 4615
Description: Located 300m from A4103.

Site Name: **Brick Close**
HSM: 32449 GR: 6180 4670
Description: By a bridleway on north side of field.

Much Dewchurch

Site Name: **Brickland**
HSM: 32361 GR: 4710 3270
Description: Lies on the B4348 and 1,250m from Tram Inn, tramway and railway.

Much Marcle

Site Name: **Brickland**
HSM: 32362 GR: 6697 3327
Description: Next to Hall Wood. It is 800m from Hellens House to the west, and Hallwood Green lies 500m to the east, tithe number 1839.

Site Name: **Brick Field Allotment**
HSM: 32444 GR: 6605 3375
Description: Located beside the A449 by Hall Farm and 800m from the village.

Site Name: **Brick Close**
HSM: 32443 GR: 6640 3345
Description: Adjacent to Hellens House. The pool on south side of the field is shared with the next entry Brick Close & Brick Close Nursery HSM 32363 for Hellens brickmaking site HSM 32363. See the section in the narrative for Hellens pp.57-59.

Site Name: **Brick Close & Brick Close Nursery**
HSM: 32363 GR: 6630 3340
Description: Next to Hellens. Sourced from the Enclosure Act 1797. See also Brick Close HSM 32443, above.

Munsley

Site Name: **Brick Clamp**
HSM: 32365 GR: 6755 3910
Description: Probably a single use site and lying beside the A438.

Site Name: **Brick Meadow**
HSM: 32366 GR: 6539 4110
Description: Probably a single use site.

Site Name: **Brick Clamp Orchard**
HSM: 32364 GR: 6352 3918
Description: Probably a single use site and to be found in detached part of parish.

Newton (Leominster)

Site Name: **Cold Oak Brickyard**
HSM: 32283 GR: 5055 5382
Description: In Hereford Record Office (E41/326 OL 365) accounts from Hampton Court estate show brick came from Cold Oak Brickyard. A house today called Cold Oak lies 1km to the west of the grid reference at Newton. See also Hampton Court Brickworks for the 19th and 20th centuries, HSM 32845 and 32849, in Hope-under-Dinmore.

Norton Canon

Site Name: **Brick Field**
HSM: 32454 GR: 3810 4970
Description: Sited 200m south of Hyatt Sarnesfield, it could have served the farm.

Site Name: **Brickfield**
HSM: 32367 GR: 3830 4680
Description: This is an isolated field about 500m south-east of Upper Norton.

Site Name: **Brick Kiln Piece**
HSM: 32368 GR: 3820 5020
Description: Sited 500m north-east of Hyatt Sarnesfield, it could have served the farm.

Ocle Pychard

Site Name: **Brick Lamp**
HSM: 32372 GR: 5845 4515
Description: Lies between Howberry and Kymin. (Brick clamp?).

Site Name: **Brick Lamp**
HSM: 32370 GR: 5977 4650
Description: Located next to the A417. (Brick clamp?).

Site Name: **Brick Lamp Piece**
HSM: 32371 GR: 5905 4641
Description: Found beside a minor road. (Brick clamp?).

Orleton

Site Name: **Brick Kiln Piece**
HSM: 32373 GR: 5040 6640
Description: Pond on the north-western boundary of the field and there is a track to the canal 300m away to the north-west.

Pembridge

Site Name: **Brick Close**
HSM: 32459 GR: 3430 5860
Description: Located 600m north-west of Sheriffs Farm. Note also the ponds at south-east corner by the dismantled railway.

Site Name: **Brick Kiln Piece**
HSM: 32376 GR: 3815 6125
Description: Near to a track leading to the B4362.

Site Name: **Brickyard Field**
HSM: 32375 GR: 3753 5749
Description: In a field by the A44. Note especially the Brickyard Cottages 500m to the west, and the only concrete remains to identify this site.

Site Name: **Brick Close**
HSM: 32374 GR: 3795 5840

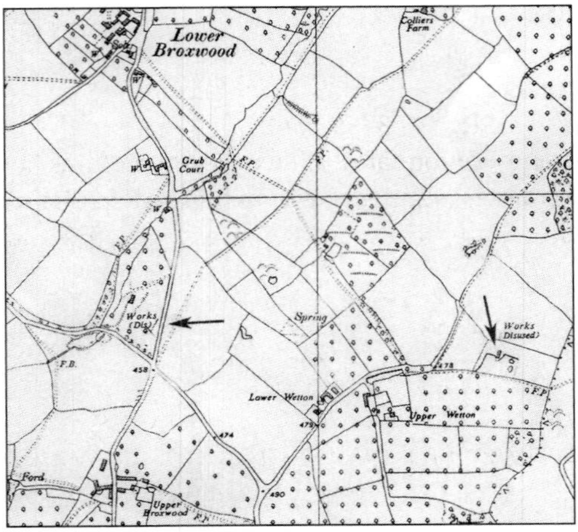

Map SO53SW shows both the Lower Broxwood and Upper Wetton brickworks

Description: Lies in the angle formed by the dismantled railway and the Rowe Ditch.

Site Name: **Lower Broxwood Brickworks**
HSM: 31578
GR: 3661 5382
Description: The brickworks are in the north corner of the crossroads 700m south of Lower Broxwood and recorded as disused on the 1964 SO35SE map. In 1889 two large and one small buildings were shown, with a large (1ha) pit which remains visible today. There is a claypit across the road from the site.

Site Name: **Bricklay Coppice**
HSM: 32457 GR: 3500 5520
Description: Just off the A480 1km north of Woonton.

Site Name: **Brick House Orchard**
HSM: 32460 GR: 3685 5960
Description: Situated 300m north-west of Court of Noke.

Site Name: **Brick Kiln Piece**
HSM: 32458 GR: 3605 5520
Description: Field name Brick Kiln Piece (tithe field no 1408) is 800m south-east of Old Moorcourt. This is the brick kiln listed at Moor Court, Kington that made 290,000 bricks per annum. The land was owned by the Rev. J. Davies and the manufacturer was Humphrey Poole. Davies is listed as the landowner of 1840 in the 1840 tithe map.

Site Name: **Brickworks**
HSM: 30553 GR: 3806 5743
Description: An undated sales map shows one pit, tithe field no 669, area 2.579 acres, and a pump adjacent on the east, also in a pit about one third the size of the first. It was marked as Old Brick Works and Pump respectively and lay about 120m to the west of Pitfield Farm. The first edition of the OS map of 1886 marks the pit as Brick Works and Clay Pits respectively, no pump is marked. The 1962 1:10560 map shows the pits as Old Works.

Site Name: **Brick Meadow**
HSM: 32456 GR: 3880 5920
Description: Flat field 1km north of village.

Site Name: **Upper Wetton Brickworks**
HSM: 31577 GR: 3739 5362
Description: A brickyard 300m east of Upper Wetton farm, shown as two buildings on the 1889 map with a claypit (0.25ha). The site is presently on a bridleway and planted with trees around the north and east sides. On a visit in 2003 nothing was found of structures.

Site Name: **2 Clay Pits**
HSM: 41080, 41079 GR: 3812 5744, 3807 5746
Description: Although not positively identified, these clay pits lie close to the Brickworks HSM 30553. Identified from the OS first edition map.

PENCOMBE WITH GRENDON WARREN

Site Name: **Brickworks**
HSM: 31045 GR: 5890 5290
Description: A series of hollow ways, rectangular enclosures (at least four) and roughly 2ha in size, marked by banks and ditches, and linked by tracks can be seen as earthworks on the east-facing hillside. At the time of the tithe this was a small croft settlement around Stone Hill (HSM 31044) with a school. By the first edition the settlement had completely disappeared and all that is marked on the map is a brick kiln and a kiln. Records show works were in production until 1882. It is not possible to say clearly whether it was these works, but as it was at Pencombe Brickyard it seems very likely it was these.[81]

Site Name: **Hegdon Hill Brick Kiln**
HSM: 31576 GR: 5842 5366
Description: In 1891 (survey 1885) little remained of the kiln or clay pits recorded at GR 5844 5361. They still showed in 1965 on map SO55SE. In 2003 there was nothing to be seen at all though there was a wide brick scatter.

Site Name: **Clay Pit**
HSM: 41138 GR: 5844 5361
Description: Although we have not positively identified a brick activity, it could be a satellite pit for Hegdon Hill Brick Kiln HSM 31576. Sourced from the first edition OS map.

PETERCHURCH

Site Name: **2 Clay Pits**
HSM: 4126, 41315 GR: 3068 4036, 3614 3866
Description: We have been unable to positively identify any brick activity nearby. Sourced from the first edition OS map.

Pipe and Lyde

Site Name: **Lower Lyde Brick Kiln**

HSM: 31562 GR: 5164 4405

Description: In 1886 claypits (quarries) and a kiln in an orchard were shown on the map. These had been cleared by the time of the publication of the 1:25000 OS Explorer map 202. It was probably used for Lower Lyde Farm 200m to the north or for Lyde Court, 400m to the west. A site visit was made in 2003 when a very slight brick scatter was observed, however there were a number of whole bricks in the northern ditch.

Pipe Aston

Site Name: **Part of Brick Kiln Piece**

HSM: 32288 GR: 4785 7330

Description: Sited unusually high on the now forested limestone hills west of Ludlow. Pipe Aston is better known for its pipeclay and tobacco clay pipe manufacture.

Preston Wynne

Site Name: **Brick House, possible brickworks**

HSM: 31618 GR: 5565 4710

Description: The site is called Brick House on the first edition map. The map also shows ponds in the vicinity that may be connected with digging for clay or other brick making processes. An aerial photograph shows earthworks. While these have been interpreted as house platforms (see HSM 31617), those to the rear (north) of Brick House may be brick working remains.

Putley

Site Name: **Brick Close Orchard**

HSM: 18561 GR: 6472 3652

Description: Located 300m north-west of Hatsford Farm.

Site Name: **Clay Pit**

HSM: 41379 GR: 6536 3743

Description: We have been unable to positively identify any brick activity nearby. Sourced from the first edition OS map.

Ross on Wye

Site Name: **Brick Meadow**

HSM: 32377 GR: 6219 2530

Description: Below Rudhall Pools. See also Weston under Penyard, as this site lay across the parish boundary.

Site Name: **Brickyard Meadow**

HSM: 32380 GR: 6072 2490

Description: One brickworks called the 'Atlas Works' in Ross was described in the Geological Memoirs of Gt. Britain and this may well be it. The source material was Blue Marl, the owner was Mary Edwards, and the manufacturer was Thomas Shott. The works produced 200,000 bricks and 60,000 tiles per annum. Now under housing.

Site Name: **Brick Meadow**
HSM: 32378 GR: 6035 2550
Description: Location deduced from tithe map, and the grid reference is only approximate as it is now under a warehouse.

Site Name: **Brickyard Meadow**
HSM: 32379 GR: 5969 2460
Description: Also described as 'Brickyard etc. and a Carpenter's Shop', and situated below Prospect Terrace in Greytree Road. This may have been the brickworks listed in the mineral statistics produced by the Geological Survey, one of 25 in the county. It was described as being a brickworks on alluvial clay owned by John Edwards Esq.

Rowelstone

Site Name: **Clay Pit**
HSM: 41349 GR: 3605 2830
Description: We have been unable to positively identify any brick activity nearby. Sourced from the first edition OS map.

St. Devereux

Site Name: **Crizeley Brickworks**
HSM: 31559 GR: 4427 3207
Description: This small works is sited about 150m north of the A465 where there is a large claypit and there was a kiln. A house is on site today, while the rest is wooded. Criseley Brick and Tile Manufacturing Co., with George Morgan as Manager, was listed in the Kelly's 1891 *Directory*. Until about 20 years ago there was machinery still on the site.[82]

The site of the Crizeley Brickworks claypits is now partially planted with poplars

Site Name: **Brick Close**
HSM: 32465 GR: 4330 3135
Description: Situated beside the A465. It was probably a single use site.

Site Name: **Brick Close**
HSM: 32466 GR: 4590 3240
Description: Found 200m from the A465. It was probably a single use site.

St. Weonards

Site Name: **Brick Ground**
HSM: 32381 GR: 4841 2380
Description: On opposite sides of the Garrett Brook.

Sarnesfield

Site Name: **The Batch Brickworks**
HSM: 31567 GR: 3694 5142
Description: This brickworks, clay pit and kiln were described as disused on the 1891 map. Now in a mixed wood with a track passing a building and a small industrial complex.

Site Name: **Brick Close**
HSM: 32313 GR: 3765 5350
Description: Located beside the drive to Nunsland Farm. Close by are tithe field nos 1248 Brick Close Meadow, 1246 Brick Close Meadow, 1241 Brick Meadow. This area is now in Sarnesfield but at the time of working was a detached part of Eardisland.

Sellack

Site Name: **The Brick Field/The Pool Field**
HSM: 32382 GR: 5458 2550
Description: Sited by the A49 with a pool to north.

Site Name: **Brick Meadow**
HSM: 32383 GR: 5540 2545
Description: By minor road and green lane both leading to the A49.

Sollers Hope

Site Name: **Brick Kiln Orchard**
HSM: 32385 GR: 6142 3320
Description: About 100m north-west of the church.

Site Name: **Brick Wood Ground**
HSM: 32464 GR: 6115 3310
Description: About 100m south-west of the church.

Stanford Bishop

Site Name: **Wofferwood Common Brickworks**

HSM: 31575 GR: 6955 5168

Description: The site is on Wofferwood Common which was fenced off in 1889 and the clay pits may have been across the road. The land has since been restored to Wofferwood Common. This may have been the site listed in Kelly's *Directory* of 1891 owned by Thomas Patterson and Son on Woodford (*sic*) Common. It was grazing land in 2003.

Stapleton and Willey

Site Name: **Brick House Orchard & Buildings**

HSM: 32387 GR: 3175 6480

Description: This is sited about 600m north of Presteigne on the road to Stapleton, and no doubt served the Presteigne market.

Site Name: **Brick Piece and Crabtree Field**

HSM: 32468 GR: 3200 6500

Description: This is sited about 1km north of Presteigne on the road to Stapleton, and no doubt served the Presteigne market.

Staunton on Arrow

Site Name: **Brick Piece**

HSM: 32386 GR: 3615 6164

Description: Now buried under outbuildings for Manor Farm. However, note the two small ponds to the east and the large pond at Manor Farm which may all have started as clay pits.

Stoke Bliss

Site Name: **Brick Kiln Meadow/Ick Field**

HSM: 32388 GR: 6473 6309

Description: By Hurst Farm. The tithe fields are: Brick Kiln number 317, and Ick Field, which is adjacent, number 316.

Stoke Lacy

Site Name: **Clay Pit**

HSM: 41140 GR: 6277 5053

Description: We have been unable to positively identify a brick activity nearby. Sourced from the first edition OS map.

Stoke Prior

Site Name: **Brick Close**

HSM: 32471 GR: 5475 5470

Description: Situated 400m south of Risbury village. Records show works were in production until 1882. See also Brickworks HSM 31045 in Pencombe with Grendon Warren.

Tarrington

Site Name: **Brick Kiln**
HSM: 18222 GR: 6121 4064
Description: North-west of The Lays Farm.

Tedstone Delamere

Site Name: **Tedstone Wafer Brickworks**
HSM: 31582 GR: 6830 5893
Description: Brickworks and claypits on the south side of the Tedstone Wafre to Tedstone Delamere Road. There are no other references to the clay pits. These are probably the brickworks listed in the 1891 Kelly's *Directory* called the Prince Albert works. Only traces remaining are a brick building, otherwise the site (in 2003) was a farm.

Site Name: **2 Clay Pit**s
HSM: 41113, 26429 GR: 6767 5893, 7102 5930
Description: We have not been able to positively identify a brick activity in the vicinity. 41113 may have been associated with Tedstone Wafer Brickworks HSM 31582. Sourced from the first edition OS map.

Tedstone Wafer

Site Name: **Clay Pit**
HSM: 41115 GR: 6731 5905
Description: We have not been able to positively identify a brick activity in the vicinity.

Thornbury

Site Name: **Clay Pit, Old Clay Pit**
HSM: 41067, 22194 GR: 6362 6051, 6355 6050
Description: We have not been able to positively identify a brick activity in the vicinity. Sourced from the first edition OS map.

Titley

Site Name: **Brickyard**
HSM: 32461 GR: 3260 5895
Description: Located beside the B4355.

Site Name: **Brickfield**
HSM: 32391 GR: 3282 5985
Description: Near the village centre.

Treville

Site Name: **Brickshed Wood and Kiln Dingle**
HSM: 31558 GR: 4317 3281
Description: Now (and in 1887) woods 500m north-west of The Valletts. The names seem to be the only clues left.

Ullingswick

Site Name: **Brick Clamp Piece**
HSM: 32390 GR: 5815 4982
Description: Between the A417 and a brook, and probably a single use site.

Site Name: **Brick Clamp Meadow**
HSM: 32389 GR: 5875 5041
Description: Located by Lowdy Farm, and probably a single use site.

Upper Sapey

Site Name: **Brick Piece or Little Criftin**
HSM: 32394 GR: 6755 6445
Description: Above village, and probably a single use site.

Site Name: **Brick Kiln Meadow**
HSM: 32395 GR: 7080 6370
Description: Lies on the county boundary with Worcestershire and 600m east of the village.

Upton Bishop

Site Name: **Brickyard Five Acres**
HSM: 32396 GR: 6348 2890
Description: Located next to a farm and probably a single use site.

Vowchurch and Turnastone

Site Name: **Brick Close**
HSM: 32392 GR: 3679 3650
Description: At junction of B4348 and B4347. Marked as a kiln on the first edition OS map with buildings and a small circular building, no doubt a kiln. Today there is a large bank at the east edge of the field that looks as though it marks the edge of a reservoir. It has a tree growing on it over 100 years old and it is marked on the first edition map. This may have been a clay pit and mixing area for the brick works.

Site Name: **Lower/Upper Brick Meadow**
HSM: 32393 GR: 3604 3621
Description: Located to the south of the village.

Walford by Ross

Site Name: **Brick Kiln Cottage**
HSM: 35557 GR: 5960 1975
Description: A brickworks at Forest Green is marked on the first edition OS map on the north side of the triangle of roads. On the 1964 map a house is shown with earthworks adjacent, presumably the quarry pit. A site visit (2002) revealed the remnants of a pond and some pits being filled with rubble. Brick Kiln Cottage was not found, the nearby Peartree Cottage is mostly stone built.

Site Name: **Brickyard Piece**
HSM: 35558 GR: 6070 1980
Description: Two adjacent fields at The Dam, Kiln Green, are called Brickyard Piece and Brickyard Field on the tithe map. On the OS first edition the fields are called Old Sandpits and on the 1964 map the earthworks and an adjacent sandpit are shown. A site visit revealed two small wooded areas in a pasture area on a hill. These were interpreted as the clay pit and kiln.

Site Name: **Brickyard**
HSM: 2432 GR: 5890 2250
Description: South-east of Arbour Hill are two small enclosures 'The Brickyard' to the north, and 'Claypit House etc.' to south. There are buildings on the south-east and north-east sides. This is probably the brickworks listed in the mineral statistics of the Geological Association of 1858. It was on alluvial clay, and the manufacturer was N. Carter. Industrial farm buildings now cover the site.

Site Name: **Brickyard Cottages**
HSM: 35556 GR: 5830 2250
Description: Brickyard Cottages are marked on the north side of the Hill Court Road, Hom Green, between Hom Green and Ross. A site visit in 2003 revealed a row of three (probably four originally) pebble-dashed cottages but no evidence of brickworks.

Site Name: **Upper Brick Pool Field**
HSM: 35559 GR: 6040 2020
Description: On the tithe map a field is marked Upper Brickpool Field, and Little Howle is adjacent to the road on the west side. On the 1885 OS map these are no longer present. On the 1964 OS map there is a Brickpool Cottage just set back from the road. A site visit in 2003 revealed a pool and earthworks in the field adjacent to the tithe Brickpool Field. The field itself is level.

Site Name: **The Brickyard Piece/Field**
HSM: 32397 GR: 6070 2000
Description: Between Rose Farm and Great Howle. These are two fields one called Brickyard Piece and the other The Brickyard Field.

Site Name: **Upper/Lower Brick Pool Field**
HSM: 32469, 32470 GR: 6030 2020
Description: Sited 150m north of Rose Farm.

Walford, Letton & Newtown
Site Name: **Clay Pit Brick Kiln**
HSM: 41021 GR: 3798 7157
Description: Sited in the middle of a field about 1km north of Letton hamlet. Sourced from the first edition OS map. It has an interesting name tying the clay with brickmaking.

Wellington

Site Name: **Brick Kiln**

HSM: 18901 GR: 4980 4970

Description: On the southern slopes above the road to Dinmore Manor. Chancehill Wood, a good source of fuel, lies further up the slope.

Site Name: **Brick Kiln Piece, Lower Brick Kiln Piece**

HSM: 11151 GR: 5020 5040

Description: Vague earthworks are identified in an orchard east and south of Burghope Court farmhouse.[83]

Weobley

Site Name: **Brick Meadow**

HSM: 32398 GR: 3847 4952

Description: This site lies 2km from Weobley and about 300m to the west of the B4230, near Shoals Bank Farm.

Site Name: **Brick Close Orchard**

HSM: 32476 GR: 4060 5160

Description: The orchard is now built up and lies 200m from the village centre on the minor road to Dilwyn.

Westhide

Site Name: **Clay Pit**

HSM: 41375 GR: 5802 4390

Description: Still extant as a pond in Westhide Wood. We have found no brick activity in the locality. Sourced from the first edition OS map.

Weston Beggard

Site Name: **Shucknall Brickworks**

HSM: 32483 GR: 5800 4200

Description: In the 1891 Kelly's *Directory* Alexander James is listed as the owner/worker of a brick works at Shucknall, Weston Beggard. Nothing as yet is known about this brickworks except bricks were produced with a good stamp 'SHUCKNALL WORKS NA HEREFORD' and used in the local cottages. One brick has been donated to the HSM. The fabric is bright red, the texture is fine grained with few stones and has the appearance of being made in the 19th or 20th century. It was not shown on tithe map or the OS first edition map though many quarries are marked on the latter. Most of these would be stone quarries, however, utilising the Shucknall outcrop of Silurian limestones.

Weston under Penyard

Site Name: **Part of Brick Meadow**

HSM: 32399 GR: 6215 2535

Description: See also Ross-on-Wye field No 496, HSM 32377, as this site lay across the parish boundary.

Whitbourne

Site Name: **Piece adjoining Brickfield Close**
HSM: 32473 GR: 7025 5745
Description: Located 200m west of Rosemore.

Site Name: **Brick Orchard**
HSM: 32474 GR: 7085 5725
Description: This is still an orchard. There is a pond on the south which may well be the associated clay pit.

Site Name: **Brick Ground**
HSM: 32475 GR: 7185 5665
Description: Now a housing estate.

Site Name: **Clay pit**
HSM: 26453 GR: 7340 5850
Description: This is too far away from the other entries for the parish to have been associated with them, and it is sited near the River Teme.

Wigmore

Site Name: **Brick Kiln Piece/Coppice/Coppice**
HSM: 24490 GR: 3900 6875
Description: Brick Kiln Coppice still exits in the woods above this very remote site (in terms of carriage of bricks to a wider market), though there are eight buildings within 750m.

Site Name: **3 Clay pits, Deerfold Clay Pit**
HSM: 41028, 41026, 36491, 31586
GR: 3952 6763, 3893 6718, 3895 6715, 3880 6720
Description: The clay pits in this area are unlikely to be for brickmaking, more likely for marling fields. The Deerfold Clay Pit HSM 31586, is a large pit about 50m by 25m. Sourced from the first edition OS map.

Willersley and Winforton

Site Name: **Brick Close**
HSM: 32400 GR: 2911 4796
Description: Next to Winforton Wood Farm, and most likely served the farm.

Site Name: **Brick House Orchard**
HSM: 32462 GR: 2990 4775
Description: This is located about 600m north of the village by a trackway. This together with HSM 32463 may have served the village.

Site Name: **Brick Meadow**
HSM: 32463 GR: 2920 4700
Description: Beside the A438 just to the west of the village. This together with HSM 32462 may have served the village.

WITHINGTON

Site Name: **Brick Kiln**
HSM: 18897 GR: 5495 4285
Description: In Brick Clamp Meadow and east of Eau Withington.

Site Name: **Godwin Tile Works (2nd), from 1861**
HSM: 11863 GR: 5650 4230
Description: With expansion of business, a new factory was built next to Withington railway station in 1861 and many of the mid-Victorian buildings still stood when the premises were occupied by Hereford Whitestone Tiles Ltd. in the 20th century.

The original tile business moved here from Lugwardine, and in 1870 William and Henry Godwin broke the partnership, Henry opening the Victoria Tile Works. In 1858 William Godwin made encaustic and mosaic tiles, bricks, draining tiles and pipes, and 850,000 bricks *per annum* here. The land was owned by C.A. Mason Esq. Godwin tiles can be seen in the chancel of Pencombe church.

The site is now Whitestone Business Park and I. & K. Brown Antiques. See also Godwin Tile Works (1st), from 1849, HSM 11862, under Lugwardine parish. For a full description of the business, see the box on the Godwin Family on p.48.

WOOLHOPE

Site Name: **Brick Lamp Meadow**
HSM: 32404 GR: 6070 3520
Description: Located 750m south of the village. (Brick clamp?).

Site Name: **Brick Close Croft**
HSM: 32477 GR: 6130 3450
Description: On a hill slope overlooking Alford's Mill to the west and Tack Wood to the south.

Site Name: **Bricky Croft**
HSM: 32405 GR: 5990 3565
Description: Located 1km south-west of the village.

Site Name: **Brick Lamp**
HSM: 32478 GR: 6160 3380
Description: Situated in a field 400m north of Ashfield. (Brick clamp?).

Site Name: **Brick Less**
HSM: 32479 GR: 6480 6340
Description: Lies 300m from Hatsford Farm. (Brick Leas?).

Site Name: **Brick Kiln Orchard**
HSM: 32402 GR: 6110 3782
Description: Below Canwood Farm.

Site Name: **Brick Lamp Orchard**
HSM: 32480 GR: 6440 3805
Description: Surrounded by many farms and 400m west of the Brainge. (Brick clamp?).

Site Name: **Brick Kiln Meadow**
HSM: 32403 GR: 6251 3640
Description: Close to minor road above Winslow Mill.

YARKHILL

Site Name: **Newtown Brickworks**
HSM: 31563 GR: 6176 4500
Description: The brickworks were disused in 1886. They were situated 150m north of the Newtown cross roads. The buildings then were extant about 100m west of the Leominster road. A visit in 2003 found a flat field.

YAZOR

Site Name: **Brick Kiln**
HSM: 18530 GR: 4070 4670
Description: Lies on a hill slope to the west of Jubilee Bridge.

Site Name: **Brick Meadow**
HSM: 32406 GR: 3876 4769
Description: Next to Bunns Lane Farm.

Site Name: **Clay pits at Yazor**
HSM: 31568 GR: 3944 4741
Description: Recorded as a shrunken settlement on the HSM. On the tithe map nine dwellings are shown, in 1891 two buildings were recorded and on the OS Explorer 201 only one appears. The present cottage is 400m north up a trackway from Upperton Farm.

Bibliography

Anon. 1851 Lascelles' *Directory*

Anon. 1879-1891 Kelley's *Directories*

Anon. 1892 *Historical Description of Hereford*

Attfield, C.E. 1999 'Some benefactors of Herefordshire in bygone days ' *Transactions of the Woolhope Naturalists' Field Club* Vol.49, 1999, Part III

Bailey BA. 1962 'William Brydges and the re-building of Tyberton Church', *Transactions of the Woolhope Naturalists' Field Club*, Vol.37

Ballard, Stephen 1999 *Colwall collection: memories of a Herefordshire village* Cora Weaver, Malvern

Beswick M. 1993 *Brickmaking in Sussex.* Middleton Press, Midhurst

Boynton, John 1995 *Rails Through the Hills* Mid England Books, Kidderminster

Brian, Anthea 1972 'A regional survey of brick bonding in England and Wales' *Vernacular Architecture* 3, 11-16

Brian, Anthea 1980 'The distribution of brick bonds in England up to 1800' *Vernacular Architecture* 11, 11-16

Brodribb, Gerald 1987 *Roman Brick and Tile* Alan Sutton Publishing, Gloucester

Bull, Dr. 1869 'Bringewood Forge' *Transactions of the Woolhope Naturalists' Field Club* 54-57

Campbell, Bruce M S, Kenneth C Bartley and John P Power 1996 The Demesne-Farming Systems of Post-Black Death England: A Classification *The Agricultural History Review* Vol 44 Part II pp131-179

Clark, Gregory 2005 The Condition of the Working Class in England, 1209–2004 *Journal of Political Economy*, volume 113 (2005), pages 1307–1340 and papers on the Internet.

Clarke, S., R.,Jackson and P. Jackson 1985a, 'A Malvernian Floor Tile Kiln at Monmouth' *Medieval Ceramics* 16

Clarke, S, R Jackson and P Jackson 1985b, 'Post-Medieval Potteries in North Gwent' *Medieval and Later Pottery in Wales* 8

Coningsby 1725 Map of Marden Manor, HRO

Connolly, Andrew 2003 *Life in the Victorian brickyards of Flinshire and Denbighshire* Gwasg Carreg Gwalch, Llanrwst

Cullingworth, J B and V Nadin 1994 *Town and country planning in Britain* Routledge, London

Dixon, J.B. and Weed, S.B. (eds) 1989 *Minerals in Soil Environments* 2nd edition Soil Science Society of America

Duncumb, John 1805 *Collections towards the history and antiquities of the County of Hereford*, with additions by William Henry Cooke/ Published by Merton Priory Press, Cardiff in 1997.

Dyer, C. 2002 *Making a Living in the Middle Ages. The People of Britain 850-1520* Yale University Press London

Eames, Elizabeth 1992 *Medieval Craftsmen; English Tilers* British Museum Press, London

Farmer, David L 1988 Prices and Wages in Hallam, pp 716-779

Greene B. 1981 *Godwin and Hereford* Anon Trade Magazine
Hallam, H E (ed) 1988 *The Agrarian History of England and* Wales Vol II 1042-1350 (general editor, Joan Thirsk) Cambridge University Press
Hammond M. 1981 *Bricks and Brickmaking,* Shire, Princes Risborough
Hillaby, J. 2003 *St Katherine's Hospital, Ledbury 1230-1547* Ledbury and District Society Trust Ltd with Logaston Press
Hunt R. 1860 *Memoirs of the Geological Survey of Great Britain. Mineral Statistics. Part 2* HMSO, London
Hurst, Derek 2002 *Castle Moat* Leominster Historical Society
Jakeman and Carver. *c* 1912 *A Short History of Hereford* William Collins
Jakeman and Carver. *c* 1914 *Historical Landmarks of Hereford,* William Collins
James, Duncan 2004 An analysis of fourteen post medieval buildings in Pembridge, Herefordshire, unpublished manuscript
James, Duncan 2005 *An Analysis of the Historic Fabric of Thirty Buildings in Eardisley Herefordshire* a report prepared for Eardisley History Group; unpublished manuscript
Lethaby, W.R. 1906 *Westminster Abbey and the King's Craftsmen* Duckworth and Co, London
Lloyd, Nathaniel 1983 *A history of English Brickwork*, with a forward by Sir Edwin Lutyens H. Greville Mongomery, London published by Antique Collectors' Club, Woodbridge
Lovelace, David 2004 *A Study of Tudor Bringewood* a report for the Forestry Commission, at www.smr.herefordshire.gov.uk
Luiten van Zanden. Jan 2003 *Prices and wages and the cost of living in the western part of the Netherlands, 1450-1800* published on the internet
Lynch, Gerard 1994 *Brickwork, History, Technology and Practice* Vol 1 Donhead Madison, Wisconsin, USA
Marshall, George, 1931 'A Pottery Site at Pembridge' *Transactions of the Woolhope Naturalists' Field Club,* pp.77-79
McWhirr, Alan 1979 *Tile-Kilns in Roman Britain* in McWhirr, pp.97-191
McWhirr, Alan (ed) 1979 *Roman Brick and Tile* BAR International Series 68
Munthe, Malcolm 1957 *Hellens. The Story of a Herefordshire Manor*
Newman, John n.d. *Welcome to the Town Hall* unpublished notes held in the Town Hall
Postle, David 1985 *From Ledbury to Gloucester by rail: as seen throuth the eyes of the driver and fireman of the last passenger train on the branch* 11 July 1959 Amber Graphics, Ledbury
Postle, David n.d. *A Glimpse of Old Ledbury* Amber Graphics, Ledbury
RCHM Royal Commission of Historical Monuments 1931-1934 *Inventory of Monuments, Herefordshire*
Renton, Charles 1999 *The Story of Herefordshire's Hospitals* Logaston Press
Roberts, Graham 2001 *The Shaping of Modern Hereford* Logaston Press
Rogers, J.ET. 1882 *Agricultural Prices and Wages in England 1259 - 1783. Volume 3 1401 - 1582* Clarendon Press, Oxford
Rydin, Yvonne 1998 *Urban Environmental Planning In The UK.* Macmillan, Basingstoke, pp.6-10

Salzman, L.F. 1952 *Building in England Down to 1540* Clarendon Press, Oxford
Schulze, D.G. 1989 A*n Introduction to Soil Mineralogy* in Dixon and Weed
Simpson, Helen J. 1997 *The day the trains came : the Herefordshire railways : the people who built them and who rejoiced when they arrived* Leominster, Gracewing
Smith, T.P. 1985 *The Medieval Brickmaking Industry in England 1400-1450* BAR British Series 138
Smith, William H. 1998 *Herefordshire Railways* Sutton, Stroud
Vince, Alan G. 1984 *The Medieval Ceramic Industry Of The Severn Valley* Unpublished PhD Thesis, University of Southampton, 1984
Waters, P.L. 1963 'A Romano-British tile kiln at Upper Sandlin Farm, Leigh Sinton, Worcestershire' *Transactions of the Worcestershire Archaeology Society* XL, pp.1-5
Whitehead, D. 1981 'Brick and tile making in the woodlands of the West Midlands in the 16th and 17th centuries' *Vernacular Architecture* 12, pp.42-47
Williams, Phyllis, 1987 *Bromyard Minster, Manor and Town*, Bringsty Publishers
Woodforde, John 1976 *Brick to Build a House* London Brick Publication
Woolley L. 1952. *Ur of the Chaldees* Penguin, Harmondsworth

References

1. Schulze 1989
2. Woolley 1952
3. Waters 1963
4. Herefordshire Museums, pers comm
5. McWhirr 1979
6. Eames 1992
7. Vince 1984
8. Andy Boucher, Arcaheological Investigations Ltd., pers comm.
9. Vince 1984
10. Marge Feryok, Archenfield Archaeology, pers comm.
11. Clarke *et al* 1985a
12. Bevan and Hodges, builders, pers comm
13. David Lovelace, pers. comm. Pembridge from Marshall 1931, Downton from Bull 1869
14. Whitehead 1981
15. Vince 1984
16. Hurst 2002
17. Clarke *et al* 1985b
18. Duncumb 1805, p. 415
19. HRO P94/6; and Mocas records of July 1777 state 1,050 gorst faggots cost £1 1s (HRO J56/IV/2)
20. Prices and wages have been taken from Rogers 1882, Lethaby 1906, Lloyd 1983, Salzman 1952

21. HRO B34/29
22. Farmer (1988) used and corrected Rogers' (1882) figures with other data to construct a cross-country ten-year grain price average for the medieval period. It shows a tonne of wheat varied in price between 343d/t and 469d/t between 1160 and 1520 and that of barley between 265d/t and 365d/t. Prices fluctuated considerably (Campbell *et al* 1996, p.27) but were highest in the 14th century and slightly higher in the 16th century than the 12th (Dyer 2002, p.230)
23. Clark 2005, Luiten van Zanden 2003
24. Lloyd 1983
25. Woodforde 1976
26. Lynch 1994, Baggeridge Bricks pers comm
27. Lynch 1994, *Hereford Times* property section; *Financial Times* 12/3/2005
28. Hampton Court day book HRO E41, Clark 2005, Connolly 2003
29. Connolly 2003
30. David Whitehead card index based on research in HRO
31. The Church of Jesus Christ of Latter Day Saints (The Mormons), have collated the British 1881 Census which they published on 25 CD-ROMs. This is the most comprehensive source of information about British people from the last 20 years of the 19th century. In addition to this source we have used records collected by David Whitehead from the Hereford and Cathedral Record Offices and elsewhere from various sources of people employed in the brick trade. The 1871, 1891 and 1901 censuses are available on the internet at a charge, but we found that to trace such a small band as Herefordshire brickmakers from this source would have been both excessively time consuming and prohibitively expensive. Nevertheless, it is surprising just how many workers and managers can be positively identified with particular brickworks from the 1881 census, particularly as most addresses given in the census are usually quite vague in modern terms. Those who gave their addresses at a brickyard or its immediate vicinity, not only indicate that they worked there but also that the works were in operation at the time. These certainties occur at Burghill, Ewyas Harold, Hereford, Holmer, Kentchurch, Kington and district, Leintwardine, Leominster, Laysters, Lugwardine. Lyonshall, Mansell Lacy, Pembridge, St. Devereux, Sarnesfield, and Tupsley. Against this, it is quite obvious that not all managers, and certainly fewer workers are named. In the following lists so few are mentioned that it would have been impossible to run the businesses concerned. We can only assume that some brickworkers called themselves labourers or farmers or some other trade because they saw brickmaking only as a minor part of their working lives, or that they were missed altogether from the census.
32. Based on Lloyd 1983 and authors' own measurements
33. Parliamentary Papers, 1836 and 1847 consulted at the British Library
34. Parliamentary Papers 1847 consulted at the British Library
35. Greene 1981
36. Anon. 1851; Anon 1879-1891, P. Klein pers. comm.
37. Munthe 1957; English Heritage
38. Coningsby 1725; English Heritage

39. RCHM 1931; Attfield 1999
40. English Heritage
41. Bailey 1962
42. English Heritage
43. English Heritage; *Hereford Times* 15 June 1861; HRO photographs; Postle 1985; Boynton 1995; Simpson 1997; Smith 1998
44. Roberts 2001; Newman n.d.; Jakeman and Carver *c.*1912
45. Jakeman and Carver *c.*1914
46. Roberts 2001; Cullingworth and Nadin 1994; Rydin 1998; Tim Reed pers comm
47. John Edmonds pers. comm.
48. Avenbury Tithe Award HRO
49. *Herefordshire Archaeological News* 68
50. Noel Knight pers. comm..
51. Phyllis Williams pers. comm.
52. Glen R. Davis pers. comm.
53. Brian Wilkins pers. comm.
54. Rosemary Ballard pers. comm.
55. Additional information from Ballard 1999
56. T. Hoverd pers. comm.
57. T. Hoverd pers. comm.
58. Louise Thomas pers.comm.
59. Debbie Klein pers. comm.
60. Gill Lloyd pers. comm.
61. A.G. Cross, Lugwardine History Society, pers. comm.
62. Mr. Eckley pers. comm.
63. Roberts 2001; S.D. Berrow pers. com.; Renton 1999
64. Woolhope Archaeological Report, 1949
65. Anon. 1892
66. Bromyard & District History Society
67. Major R. Barns, pers. comm.
68. A.G. Cross, K.H. Crooks pers. comm.
69. Anon. 1892
70. George Children and James Archer pers. comm.
71. Bromyard & District History Society
72. Womens' Institute Local History Group
73. Tithe Apportionment 1842. Littlebury's *Directory* 1867–1890, Kelly's 1885, 1890
74. Mr. Poole pers. comm.
75. Major R. Barns pers. comm.
76. Norman Davies pers. comm.
77. Margaret Feryok pers. comm.
78. A.G. Cross, Lugwardine History Society, pers. comm.
79. P. Wood pers. comm.
80. Much Cowarne History Group
81. Bromyard & District History Society

82. John Edmonds pers. comm.
83. Rosamund Hickling, Sectional Report, *Transactions of the Woolhope Naturalists' Field Club*, 1970, p.174

Index